THEORY OF IONIZATION OF ATOMS BY ELECTRON IMPACT

THEORY OF IONIZATION OF ATOMS BY ELECTRON IMPACT

R.K. Peterkop

Translation edited by
D.G. Hummer
from a draft translation by Elliot Aronson

Joint Institute for Laboratory Astrophysics
National Bureau of Standards
and
University of Colorado
Boulder, Colorado 80309

COLORADO ASSOCIATED UNIVERSITY PRESS

International Standard Book Number 0-87081-105-3
Library of Congress Card Catalog Number: 77-81310
Printed in the United States of America

Originally published by Zinatne Publishing House
Riga, USSR 1975

AUTHOR'S PREFACE

The theory of atomic ionization by electron impact forms an important part of the theory of electron-atom collisions, which has undergone intensive development in the last decade. This development has been stimulated by applications to plasma physics, by the development of general collision theory, and also by growing computational possibilities which result from expanding computer technology.

There are a number of monographs devoted to the theory of electron-atom collisions and general collision theory. However, these mainly expound the theory of elastic scattering and excitation of discrete levels. For example, Drukarev [1] treats only the processes just mentioned; in the comprehensive monograph of Mott and Massey [2] the theory of ionization is presented in the Born approximation; Wu and Omura [3] give a general statement of the problem for a reaction with a rearrangement of the particles, but make no allowance for the peculiarities associated with the long-range nature of the Coulomb forces. The recent book of Vainshtein, Sobel'man, and Yukov [4] presents tables and formulae that enable one to evaluate effective ionization cross sections, but no discussion of the theoretical methods is given.

The theory of ionization has inherent specific features that distinguish it from the theory of elastic scattering and atomic excitation. In spite of this, only a few review articles [5-7] have been devoted specifically to ionization theory, whereas there is enough material available on this subject for an extensive monograph. In the present book the main consideration is given to the work carried out at the Institute of Physics of the Latvian Academy of Sciences.

In the first chapter we set forth the main concepts involved in the formulation of the problem of atomic ionization by an electron. Chapters II-VII are devoted to particular problems in the theory of ionization. The asymptotic form of the wave function for a system of three or more charged particles with positive energies is examined in detail. The method of K harmonics is discussed, in which the problem of the continuous spectrum is reduced to a problem with a discrete set of unknown functions. Many-dimensional Coulomb wave functions are introduced and used to obtain integral representations of the ionization amplitude, the role of electron exchange is discussed, and the classical theory of Wannier on the threshold behavior of ionization cross sections is discussed, as well as the semiclassical form of this theory. Chapter VIII contains a review of methods for calculating effective ionization cross sections.

The author is grateful to M. K. Gailitis and L. L. Rabik for discussions of problems in the theory of ionization.

R. Peterkop

EDITOR'S PREFACE

Collisional ionization of atoms and ions is one of the most perplexing and fascinating of the current problems of atomic physics, and no one has done more to unravel its complexities than Dr. Raimond Peterkop. Thus, the wisdom of Professor M. J. Seaton's suggestion last spring that Dr. Peterkop's new book should be translated into English was immediately apparent. Fortunately Dr. Elliot Aronson was available to provide a draft translation. To produce the book quickly and cheaply, we chose to use photo-offset from typed copy. All was facilitated by Dr. Rolf M. Sinclair of the National Science Foundation, who gave permission to pay Dr. Aronson from the NSF Block Grant PHY76-04761 to the Joint Institute for Laboratory Astrophysics. I am also grateful to the National Bureau of Standards for permitting me to edit the translation as a part of my official duties.

In addition to the individuals already named, I am deeply grateful to Ms. Gwendy Romey for rapidly and accurately typing the camera copy and to Ms. Lorraine Volsky for editorial assistance, as well as for overseeing the typing and other aspects of production. Professor Seaton read the entire manuscript and made many suggestions for improving the accuracy, clarity, and style of the translation.

It is my great pleasure to thank the author, Dr. Raimond Peterkop, for his prompt and courteous cooperation in supplying new material to be incorporated into the book, as well as in responding to my queries and checking the camera copy. I sincerely hope that readers will experience the same pleasure in using this book that I had in working with all of the above in preparing it.

David G. Hummer
1 February 1977

CONTENTS

CHAPTER I

BASIC PRINCIPLES OF THE THEORY OF IONIZATION

§1. *Boundary Conditions*

We shall consider first the formulation of the ionization problem in the simplest case -- ionization of hydrogen atoms. The basic results for this case are valid also for the ionization of multielectron atoms, so long as the incident and ejected electrons can be regarded as moving in a given field due to the nucleus and the remaining electrons. The specific features of ionization of a two-electron atom are investigated in §23.

In the present chapter we consider the incident and atomic electrons to be distinguishable particles. (Interference effects, following from the indistinguishability of the electrons, are treated in §22.) We use atomic units and assume the mass of the nucleus to be infinite.

The Schrödinger equation for a system of two electrons in the field of the fixed nucleus of the hydrogen atom is

$$\left(-\frac{1}{2}\Delta_1 - \frac{1}{2}\Delta_2 + V - E\right)\Psi(\underset{\sim}{r}_1,\underset{\sim}{r}_2) = 0 \quad , \tag{1.1}$$

where $\underset{\sim}{r}_1$ and $\underset{\sim}{r}_2$ are the radius vectors of the electrons (we assume the origin of the coordinates to be at the nucleus), and the potential energy is

$$V = -\frac{1}{r_1} - \frac{1}{r_2} + \frac{1}{r_{12}} \quad . \tag{1.2}$$

Interactions with one of the two electrons can be neglected if it is sufficiently far away from the nucleus and from the other electron. The solution of the Schrödinger equation is then a linear combination of wave functions describing the free motion of the distant electron (in the form of a plane wave or spherically diverging wave) and the Coulomb motion of the second electron. The latter is described by a wave function of the discrete or continuous spectrum of the atom. In accordance with the physical interpretation of the collision process, one state of the atom (the initial state) will be represented by a plane wave plus a diverging wave, while the rest of the states will be represented by diverging waves alone. Thus, the boundary conditions are

$$\Psi \underset{r_1\to\infty}{\sim} e^{i\mathbf{k}_o\cdot\mathbf{r}_1}\phi_o(\mathbf{r}_2) + \frac{1}{r_1}\sum_{n\ell m}\phi_{n\ell m}(\mathbf{r}_2)f_{n\ell m}(\hat{\Omega}_1)e^{ik_n r_1} + \\ + \frac{1}{r_1}\int_{v\le\kappa}\phi(\mathbf{v},\mathbf{r}_2)f_{\mathbf{v}}(\hat{\Omega}_1)e^{ikr_1+i\eta(\mathbf{v},\mathbf{r}_1)}d\mathbf{v} \quad , \qquad (1.3)$$

$$\Psi \underset{r_2\to\infty}{\sim} \frac{1}{r_2}\sum_{n\ell m}\phi_{n\ell m}(\mathbf{r}_1)g_{n\ell m}(\hat{\Omega}_2)e^{ik_n r_2} + \\ + \frac{1}{r_2}\int_{v\le\kappa}\phi(\mathbf{v},\mathbf{r}_1)g_{\mathbf{v}}(\hat{\Omega}_2)e^{ikr_2+i\eta(\mathbf{v},\mathbf{r}_2)}d\mathbf{v} \quad , \qquad (1.4)$$

where $\hat{\Omega}_1$ and $\hat{\Omega}_2$ are the directions of the vectors $\mathbf{r}_1$ and $\mathbf{r}_2$.

The first term on the right side of the asymptotic form (1.3) describes the initial state. In (1.3) and (1.4) we are assuming that electron 2 is located in the atom before the collision.

The vector $\mathbf{k}_o$ is the wave vector of the incident electron, i.e., the initial wave vector of electron 1. The functions $\phi_{n\ell m}(\underset{\sim}{r})$ and $\phi(\underset{\sim}{v},\underset{\sim}{r})$ are wave functions of the discrete and continuous hydrogen-atom spectrum:

$$\left(\frac{1}{2}\,\Delta + \frac{1}{r} + \varepsilon_n\right)\phi_{n\ell m}(\underset{\sim}{r}) = 0 \quad , \tag{1.5}$$

$$\left(\frac{1}{2}\,\Delta + \frac{1}{r} + \frac{v^2}{2}\right)\phi(\underset{\sim}{v},\underset{\sim}{r}) = 0 \quad . \tag{1.6}$$

In the continuous spectrum there are two sets of functions describing motion with a definite direction at infinity, which are usually [8] denoted as $\phi^{(+)}$ and $\phi^{(-)}$. The function $\phi^{(+)}$ has the asymptotic behavior "plane + diverging wave," and the function $\phi^{(-)}$, the behavior "plane + converging wave."[1] The function ϕ used here denotes $\phi^{(-)}$, which is needed to describe the final state when it lies in the continuous spectrum [8]. It will be shown later that only for this choice of ϕ will the quantities f and g have the meaning of ionization amplitudes, because the converging wave does not contribute to the asymptotic form of the integral over the continuous spectrum. The wave function $\phi^{(-)}$ for the hydrogen atom, normalized to $\delta(\underset{\sim}{v}-\underset{\sim}{v}')$, has the form

$$\phi(\underset{\sim}{v},\underset{\sim}{r}) = (2\pi)^{-3/2} e^{\pi/2v}\,\Gamma\left(1+\frac{i}{v}\right) e^{i\underset{\sim}{v}\cdot\underset{\sim}{r}}\;{}_1F_1\left(-\frac{i}{v}\,,\,1\,,\,-i(vr+\underset{\sim}{v}\cdot\underset{\sim}{r})\right) , \tag{1.7}$$

where $\Gamma(x)$ is the gamma function, and ${}_1F_1(x)$ is the confluent hypergeometric function.

The quantities $f_{n\ell m}$ and $g_{n\ell m}$ are the amplitudes for direct and exchange excitation of discrete levels. The amplitude $f_{n\ell m}(\hat{\Omega}_1)$ represents the case in which electron 1 leaves in the

[1] The obvious interpretation of the functions $\phi^{(+)}$ and $\phi^{(-)}$ is discussed in §26. We note that diverging wave means outgoing wave and converging wave means ingoing wave.

direction $\hat{\Omega}_1$ with wave number k_n, and electron 2 remains in the atomic state $\phi_{n\ell m}$. The amplitude $g_{n\ell m}(\hat{\Omega}_2)$ corresponds to the exchange process, in which the incident electron remains in the bound state $\phi_{n\ell m}$, and the electron which was originally in the atom escapes in the direction $\hat{\Omega}_2$ with wave number k_n.

The quantities $f_{\underset{\sim}{v}}(\hat{\Omega}_1)$ and $g_{\underset{\sim}{v}}(\hat{\Omega}_2)$ are the ionization amplitudes. Their physical meaning is discussed in more detail in §2.

The wave numbers k_n and k are determined from the conservation of energy:

$$k_n^2 + 2\varepsilon_n = 2E \quad , \tag{1.8}$$

$$k^2 + v^2 = 2E \quad . \tag{1.9}$$

The condition (1.9) restricts the domain of integration over $\underset{\sim}{v}$ in (1.3) and (1.4):

$$0 \le v \le \kappa \quad , \tag{1.10}$$

where

$$\kappa = \sqrt{2E} \quad . \tag{1.11}$$

The integrals over the continuous spectrum in (1.3) and (1.4) correspond to ionization. This portion of the asymptotic form differs from the terms corresponding to excitation of discrete levels in that an additional term $\eta(\underset{\sim}{v},\underset{\sim}{r}_1)$ has been introduced into the phase of the scattered wave. In the excitation of a discrete level the electron remaining in the atom screens the nucleus, so at large distances there is no Coulomb interaction between the scattered electron and the excited atom. In the case of ionization, as both electrons leave the nucleus, there is no screening, and at large distances there remains a slowly decaying Coulomb interaction between the nucleus and the electrons and between the electrons themselves. Therefore we can expect that for ionization the phase shift of the scattered electron will contain an additional term of the logarithmic type. Moreover, the correction to the phase shift may depend on the directions of the escaping electrons as well as on the absolute values of their velocities. In this regard the phase shift for ionization differs considerably from the phase shift for excitation of discrete levels of ions by electrons, for in the latter case the Coulomb interaction remains in the asymptotic domain, but the problem reduces to one involving just two bodies. For ionization we have essentially a three-particle Coulomb problem.

We can find the explicit form of η by comparing the asymptotic forms (1.3) and (1.4) with those found in Chapter II. The expression for η is given in §21, from which it is seen that η is real and depends on r through the factor $\ln r$.

§2. *Relationship Between the Ionization Amplitudes*

In analogy with the case of discrete-level excitation, the amplitudes f and g in the continuous spectrum can be called the direct and exchange ionization amplitudes. However, because in ionization both electrons escape from the nucleus, the concept of clearly distinguishable direct and exchange processes loses its usual meaning. The amplitude $g_{\underset{\sim}{v}}(\hat{\Omega}_2)$ corresponds to the process in which the incident electron leaves the ion with wave vector $\underset{\sim}{v}$, and the electron which was in the atom departs with wave vector $\underset{\sim}{k}$ of modulus determined by Eq. (1.9) and direction $\hat{\Omega}_2$. On the other hand, the same process is described by the amplitude $f_k(\hat{\Omega}_1)$, where $\hat{\Omega}_1$ is the direction of the vector $\underset{\sim}{v}$. Thus, in contrast with the excitation of discrete levels, the amplitudes f and g in the case of ionization are not independent.

In order to formulate the relationship between the amplitudes, it is convenient to introduce a more symmetrical notation for the arguments:

$$f_{\underset{\sim}{v}}(\hat{\Omega}_1) = f(\underset{\sim}{k},\underset{\sim}{v}) \quad , \tag{2.1}$$

$$g_{\underset{\sim}{v}}(\hat{\Omega}_2) = g(\underset{\sim}{k},\underset{\sim}{v}) \quad , \tag{2.2}$$

where the absolute values of the vectors $\underset{\sim}{v}$ and $\underset{\sim}{k}$ are related by Eq. (1.9), and the directions are independent. In (2.1) the direction of the vector $\underset{\sim}{k}$ is $\hat{\Omega}_1$, and in (2.2) it is $\hat{\Omega}_2$.

In accordance with the physical interpretation of the amplitudes we have

$$|g(\underset{\sim}{k},\underset{\sim}{v})| = |f(\underset{\sim}{v},\underset{\sim}{k})| \quad . \tag{2.3}$$

For a more rigorous derivation of the relation (2.3) we compare the asymptotic forms (1.3) and (1.4), when simultaneously $r_1 \to \infty$ and $r_2 \to \infty$ [9]. Then both (1.3) and (1.4) are applicable. We assume that as $t \to \infty$

$$\underset{\sim}{r}_1 = \underset{\sim}{v}_1 t \ , \ \underset{\sim}{r}_2 = \underset{\sim}{v}_2 t \quad , \tag{2.4}$$

where $\underset{\sim}{v}_1$ and $\underset{\sim}{v}_2$ are constant vectors related by

$$v_1^2 + v_2^2 = 2E \equiv \kappa^2 \quad . \tag{2.5}$$

In the asymptotic domain the quantities t, $\underset{\sim}{v}_1$, and $\underset{\sim}{v}_2$, defined in (2.4) and (2.5), can be interpreted as time and the velocities of the particles. In fact, if the particles simultaneously leave the interaction zone, the dimensions of which can be ignored in comparison with the remainder of the trajectory, then the distances and velocities of the particles are related by the classical equation of free motion.

Introducing hyperspherical coordinates

$$r = \sqrt{r_1^2 + r_2^2} \quad , \quad \alpha = \text{arctg}\,\frac{r_2}{r_1} \quad , \tag{2.6}$$

$$r_1 = r\cos\alpha \quad , \quad r_2 = r\sin\alpha \quad , \tag{2.7}$$

we obtain from (2.4) and (2.5)

$$r = \kappa t \quad , \quad \alpha = \text{arctg}\,\frac{v_2}{v_1} \quad , \tag{2.8}$$

$$\underset{\sim}{r}_1 = \frac{r}{\kappa}\underset{\sim}{v}_1 \quad , \quad \underset{\sim}{r}_2 = \frac{r}{\kappa}\underset{\sim}{v}_2 \quad . \tag{2.9}$$

It follows from Eqs. (2.4) that $r \to \infty$ for constant $\hat{\Omega}_1$, $\hat{\Omega}_2$, and α. These quantities determine a direction in a six-dimensional configuration space. Each such direction corresponds to two definite vectors $\underset{\sim}{v}_1$ and $\underset{\sim}{v}_2$ satisfying (2.5), i.e., to a certain outcome of the ionization process. The distribution over directions in configuration space describes both the angular distributions of the particles in three-dimensional space and the distribution of the particles over distances, which in the asymptotic domain [according to (2.4)] implies the distribution over velocities.

Having specified the way in which simultaneously r_1 and $r_2 \to \infty$, we consider first the asymptotic behavior of the expression (1.3). Ignoring the rapidly decaying discrete-spectrum terms, we obtain from (1.3)

$$\Psi \sim \frac{1}{r_1}\int_{v\le\kappa} \phi(\underset{\sim}{v},\underset{\sim}{r}_2) f(\underset{\sim}{k},\underset{\sim}{v}) e^{ikr_1 + i\eta(\underset{\sim}{v},\underset{\sim}{r}_1)} d\underset{\sim}{v} \quad , \tag{2.10}$$

where

$$k = \sqrt{2E - v^2} \quad , \quad \underset{\sim}{k} = \frac{k}{r_1}\underset{\sim}{r}_1 \quad . \tag{2.11}$$

For the atomic wave function in the continuous spectrum we use the asymptotic expression

$$\phi(\underset{\sim}{v}, r_2) \sim \frac{\delta(\hat{\Omega}_{\underset{\sim}{v}} - \hat{\Omega}_2)}{\sqrt{2\pi}\ ivr_2} e^{ivr_2 + (i/v)\,\ell n\, 2vr_2} +$$

$$+ \frac{f_3^*(-\cos\theta_{\underset{\sim}{v}\cdot\underset{\sim}{r}_2})}{(2\pi)^{3/2} r_2} e^{-ivr_2 - (i/v)\,\ell n\, 2vr_2} , \qquad (2.12)$$

the basis for which is given in §16. Substituting (2.12) into (2.10), we get two integrals which contain the rapidly oscillating exponentials

$$\exp\left[ikr_1 \pm ivr_2 \pm \frac{i}{v}\,\ell n\, 2vr_2 + i\eta(\underset{\sim}{v}, \underset{\sim}{r}_1)\right] . \qquad (2.13)$$

The asymptotic form of such integrals can be found by the method of stationary phase [10]. In this calculation, the slowly increasing logarithmic terms can be treated as constant quantities [11]. The stationary points are determined by the condition

$$\frac{k}{v} = \pm \frac{r_1}{r_2} \qquad (2.14)$$

The minus sign in (2.14) refers to the integral arising from the second term of (2.12). Since the quantities k, v, r_1, and r_2 cannot be negative, in this case the domain of integration does not contain any stationary points, and, consequently, the part of Eq. (2.12) containing the converging wave does not contribute to the leading term of the asymptotic expansion. The function $\phi^{(-)}$ used here has the advantage that the outgoing part of the asymptotic form contains a δ function in the angular variables, and this part gives the only contribution to the asymptotic form of the integral (2.10).

By comparing (2.4) and (2.14) and taking into account the presence of the δ function in the first term of (2.12), it is not difficult to see that for $r \to \infty$ only the neighborhood of the point

$$\underset{\sim}{v}_{st} = \underset{\sim}{v}_2 \tag{2.15}$$

contributes to the integral (2.10). It follows from (2.11) that the vector $\underset{\sim}{k}$ takes the value

$$\underset{\sim}{k}_{st} = \underset{\sim}{v}_1 \quad . \tag{2.16}$$

Equations (2.15) and (2.16) demonstrate that only that part of the integral (2.10), for which the particle velocities and distances are related by the classical equation of free motion, contributes to the asymptotic expression of the wave function. This is the mathematical basis for the physical interpretation of the relations (2.4) given previously.

Continuing, we substitute into (2.10)

$$d\underset{\sim}{v} = v^2\, dv\, d\hat{\Omega}_{\underset{\sim}{v}} \quad . \tag{2.17}$$

The slowly varying quantities, including the exponential with the logarithmically increasing phase, are taken through the integral sign in (2.10) with values corresponding to the stationary point (2.15). By means of the stationary-phase method we find for the remaining integral

$$\frac{1}{r_1} \int_0^{\kappa} e^{ikr_1 + ivr_2} dv \sim \frac{1}{r} \sqrt{\frac{2\pi\kappa}{ir}}\, e^{i\kappa r} \quad , \tag{2.18}$$

where r is defined in (2.6).

Complex quantities raised to some power are hereafter to be taken with the argument least in absolute value, i.e.,

$$(\pm i)^{\lambda} = e^{\pm i\lambda\pi/2} \quad . \tag{2.19}$$

Hence, (2.10) takes the form

$$\Psi \sim f(\underset{\sim}{v}_1,\underset{\sim}{v}_2)(-i\kappa)^{3/2} r^{-5/2} e^{i\kappa r+(i/v_2)\,\ell n\, 2v_2 r_2 + i\eta(\underset{\sim}{v}_2,\underset{\sim}{r}_1)} \quad . \tag{2.20}$$

Similarly we can find the asymptotic form of Eq. (1.4). In this case the stationary values corresponding to the condition (2.4) will be

$$\underset{\sim}{v}_{st} = \underset{\sim}{v}_1 \quad , \quad \underset{\sim}{k}_{st} = \underset{\sim}{v}_2 \quad . \tag{2.21}$$

Thus, in addition to (2.20), we obtain

$$\Psi \sim g(\underset{\sim}{v}_2,\underset{\sim}{v}_1)(-i\kappa)^{3/2} r^{-5/2} e^{i\kappa r+(i/v_1)\,\ell n\, 2v_1 r_1 + i\eta(\underset{\sim}{v}_1,\underset{\sim}{r}_2)} \quad . \tag{2.22}$$

Equating (2.20) and (2.22) and recalling that η is real, we obtain (2.3). The relationship between the amplitudes including their phases is treated in §21.

We note that (2.20) and (2.22) are divergent waves in the six-dimensional configuration space. A possible, and essentially more natural, way of choosing the boundary conditions for the ionization problem is to require the asymptotic form (2.20) to be satisfied in the region of configuration space where both electrons are far from the nucleus.

The asymptotic form (2.20) can also be written in the form

$$\Psi \sim A(\hat{\Omega}) r^{-5/2} e^{i\kappa r + i\gamma(r,\hat{\Omega})} \quad , \tag{2.23}$$

which will be used in the following chapters. Here γ is a real function, depending logarithmically on r, and $\hat{\Omega}$ denotes a direction in configuration space, i.e., it denotes the set of quantities

$$\hat{\Omega}_1, \ \hat{\Omega}_2, \ \alpha = \operatorname{arctg}\frac{r_2}{r_1} = \operatorname{arctg}\frac{v_2}{v_1} \quad . \tag{2.24}$$

Comparing (2.23) with (2.20) and (2.22), we obtain

$$\kappa^{-3/2}|A(\hat{\Omega})| = |f(\underset{\sim}{v}_1,\underset{\sim}{v}_2)| = |g(\underset{\sim}{v}_2,\underset{\sim}{v}_1)| \quad . \tag{2.25}$$

§3. *Effective Ionization Cross Section*

In order to define the effective ionization cross section, we must discuss the electron fluxes.

In a problem with two electrons we have appropriately two flux-density vectors [12]:

$$\underset{\sim}{J}_i(\underset{\sim}{r}_1,\underset{\sim}{r}_2) = \operatorname{Im}[\Psi^*(\underset{\sim}{r}_1,\underset{\sim}{r}_2)\nabla_i\Psi(\underset{\sim}{r}_1,\underset{\sim}{r}_2)] \ , \ i = 1,2 \ , \tag{3.1}$$

where Im denotes the imaginary part, and the operator ∇_i acts on the coordinates of the *ith* electron. The vector $\underset{\sim}{J}_i$ has the meaning of the flux density due to the motion of the *ith* electron for given coordinates of the other electron.

From the Schrödinger equation (1.1) it follows that flux is conserved:

$$\nabla_1\cdot\underset{\sim}{J}_1 + \nabla_2\cdot\underset{\sim}{J}_2 = 0 \quad . \tag{3.2}$$

The flux-density vectors of one electron for any location of the other electron are

$$\underset{\sim}{j}_1(\underset{\sim}{r}_1) = \int \underset{\sim}{J}_1(\underset{\sim}{r}_1,\underset{\sim}{r}_2)d\underset{\sim}{r}_2 \quad , \tag{3.3}$$

$$\underset{\sim}{j}_2(\underset{\sim}{r}_2) = \int \underset{\sim}{J}_2(\underset{\sim}{r}_1,\underset{\sim}{r}_2)d\underset{\sim}{r}_1 \quad . \tag{3.4}$$

Putting (1.3) and (1.4) into (3.3) and (3.4), we obtain the following asymptotic expressions for the fluxes of the individual electrons:

$$\mathbf{j}_1 \sim \mathbf{k}_o + \left(\frac{\mathbf{k}_o}{r_1} + \frac{(\mathbf{k}_o \cdot \mathbf{r}_1)\mathbf{r}_1}{r_1^3}\right) \mathrm{Re}\left[f_o(\hat{\Omega}_1) e^{ik_o r_1 - i\mathbf{k}_o \cdot \mathbf{r}_1}\right] +$$

$$+ \frac{\mathbf{r}_1}{r_1^3}\left[\sum_{n\ell m} k_n |f_{n\ell m}(\hat{\Omega}_1)|^2 + \int_{v \leq \kappa} k |f_{\mathbf{v}}(\hat{\Omega}_1)|^2 d\mathbf{v}\right] , \qquad (3.5)$$

$$\mathbf{j}_2 \sim \frac{\mathbf{r}_2}{r_2^3}\left[\sum_{n\ell m} k_n |g_{n\ell m}(\hat{\Omega}_2)|^2 + \int_{v \leq \kappa} k |g_{\mathbf{v}}(\hat{\Omega}_2)|^2 d\mathbf{v}\right] , \qquad (3.6)$$

where Re denotes the real part, and k_n and k are determined by Eqs. (1.8) and (1.9). In deriving Eqs. (3.5) and (3.6) we have taken into account the fact that because η increases as $\ln r_i$ for $r_i \to \infty$, and therefore $d\eta/dr_i \to 0$, the presence of η has no effect on the expressions for the fluxes.

The first term on the right side of Eq. (3.5) represents the flux of incident electrons; the second term expresses the interference between the incident and the elastically scattered electrons; the third term describes elastic scattering and excitation of discrete levels; and the fourth term corresponds to ionization. Equation (3.6) expresses the flux of electrons ejected from the atom by exchange excitation of discrete levels and by ionization.

The effective ionization cross section is equal to the ratio of the number of ions formed per unit time to the incident-electron flux density. The number of ions formed is equal to the number of incident electrons scattered during ionization, which is the same as the number of electrons ejected from the atom in this process. This number is equal to the flux through an infinitely distant sphere from that part of either vector $\mathbf{j}_1$ or $\mathbf{j}_2$ which corresponds to ionization. Thus, we obtain the following expressions for the effective ionization cross section:

$$\sigma = \frac{1}{k_o} \int_{v \le \kappa} k |f_{\underset{\sim}{v}}(\hat{\Omega}_1)|^2 d\underset{\sim}{v} d\hat{\Omega}_1 \quad , \tag{3.7}$$

$$\sigma = \frac{1}{k_o} \int_{v \le \kappa} k |g_{\underset{\sim}{v}}(\hat{\Omega}_2)|^2 d\underset{\sim}{v} d\hat{\Omega}_2 \quad . \tag{3.8}$$

Recall that

$$d\underset{\sim}{v} = v^2 dv d\hat{\Omega}_{\underset{\sim}{v}} \quad , \quad d\hat{\Omega}_i = \sin\theta_i \, d\theta_i \, d\phi_i \quad , \tag{3.9}$$

where θ_i and ϕ_i are spherical coordinates. Equations (3.7) and (3.8) are equal by virtue of relation (2.3).

Let us denote the wave vectors of the first and second electrons, respectively, by $\underset{\sim}{v}_1$ and $\underset{\sim}{v}_2$. We write the amplitudes in the form (2.1) and (2.2) and transform to an integration over energies:

$$\varepsilon_i = v_i^2/2 \ , \ d\underset{\sim}{v}_i = v_i d\varepsilon_i d\hat{\Omega}_i \ , \ i = 1,2 \quad . \tag{3.10}$$

Then (3.7) and (3.8) take the form

$$\sigma = \int_0^E \frac{v_1 v_2}{k_o} d\varepsilon_2 \int |f(\underset{\sim}{v}_1, \underset{\sim}{v}_2)|^2 d\hat{\Omega}_1 d\hat{\Omega}_2 \quad , \tag{3.11}$$

$$\sigma = \int_0^E \frac{v_1 v_2}{k_o} d\varepsilon_1 \int |g(\underset{\sim}{v}_2, \underset{\sim}{v}_1)|^2 d\hat{\Omega}_1 d\hat{\Omega}_2 \quad . \tag{3.12}$$

Taking (2.3) into account, we may write the effective ionization cross section in the more symmetrical form

$$\sigma = \int_0^{E/2} \frac{v_1 v_2}{k_o} d\varepsilon_2 \int [|f(\underset{\sim}{v}_1, \underset{\sim}{v}_2)|^2 + |g(\underset{\sim}{v}_1, \underset{\sim}{v}_2)|^2] d\hat{\Omega}_1 d\hat{\Omega}_2 \ . \tag{3.13}$$

In contrast with (3.11) and (3.12), the integration here is performed over the energy range $0 \le \varepsilon_2 \le E/2$, because the integrand in (3.13) takes account of all cases when one of the electrons has energy ε_2.

We also give an expression for the effective ionization cross section, which makes use of the amplitude $A(\hat{\Omega})$ and substitutes an integration over the quantity α for the integration over energy. From (2.5) and (2.24) we obtain

$$v_1 v_2 d\varepsilon_2 = \kappa^4 \sin^2\alpha \cos^2\alpha \, d\alpha \quad . \tag{3.14}$$

Taking into account (2.25), we can rewrite Eqs. (3.11) and (3.12) in the form

$$\sigma = \frac{\kappa}{k_o} \int |A(\hat{\Omega})|^2 d\hat{\Omega} \quad , \tag{3.15}$$

where $\hat{\Omega}$ denotes a direction in configuration space and is defined by (2.24), and the element of solid angle is of the form

$$d\hat{\Omega} = d\hat{\Omega}_1 d\hat{\Omega}_2 \sin^2\alpha \cos^2\alpha \, d\alpha \quad . \tag{3.16}$$

Equation (3.15) has an obvious geometric interpretation. It expresses the effective ionization cross section in terms of the probability flux in configuration space.

Equations (3.11)-(3.15) define the total ionization cross section. We can consider the integrand in (3.11) to be the differential cross section, corresponding to those ionization events in which the first electron is scattered into the element of solid angle $d\hat{\Omega}_1$, and the second electron is scattered into $d\hat{\Omega}_2$ with energy in the interval $d\varepsilon_2$. The energy of the first electron is $\varepsilon_1 = E - \varepsilon_2$. We can interpret the integrands of (3.12) and (3.13) analogously. This is indeed the case if in defining the amplitudes f and g in (1.3) and (1.4) we use the functions $\phi^{(-)}$. The expressions (3.11)-(3.13) for the total cross section are also valid if in defining f and g we use the functions $\phi^{(+)}$, since the change from $\phi^{(-)}$ to $\phi^{(+)}$ is a unitary transformation. However, f and g give the differential cross section directly only when $\phi^{(-)}$ is used, because only then do (2.20) and (2.22) follow from Eqs. (1.3) and (1.4).

For a stationary solution of the Schrödinger equation the differential ionization cross section is determined by the probability flux density in configuration space, following Gerjuoy [13]. Let us consider the probability flux through a small section of a remote hypersphere in the six-dimensional configuration space, using the hyperspherical coordinates (2.6).

The six-dimensional volume element is

$$d\underset{\sim}{r} = d\underset{\sim}{r}_1 d\underset{\sim}{r}_2 = r^5 dr d\hat{\Omega} \quad , \tag{3.17}$$

where $d\hat{\Omega}$ is defined by (3.16). Here $\underset{\sim}{r}$ denotes the six-dimensional radius vector, the components of which are $\underset{\sim}{r}_1$ and $\underset{\sim}{r}_2$. The surface element on a hypersphere in six-dimensional space is

$$dS = r^5 d\hat{\Omega} \quad . \tag{3.18}$$

The six-dimensional probability flux density, which we denote by $\underset{\sim}{J}$, has as its components the three-dimensional vectors $\underset{\sim}{J}_1$ and $\underset{\sim}{J}_2$. We note that the equation for the conservation of flux (3.2) can be written in the form

$$\nabla \cdot \underset{\sim}{J} = 0 \quad , \tag{3.19}$$

where $\nabla\cdot$ denotes the six-dimensional divergence operator.

The flux through the surface element on the hypersphere is

$$dN = J_r r^5 d\hat{\Omega} \quad , \tag{3.20}$$

where J_r denotes the radial component of the six-dimensional flux density. According to (2.20) the wave function describing ionization in the asymptotic region has the form of a spherically diverging wave in six-dimensional space. Thus the flux density is in the radial direction, and consequently

$$J_r = |\underset{\sim}{J}| = \sqrt{J_1^{\,2} + J_2^{\,2}} \quad . \tag{3.21}$$

However, it is simpler to find J_r from the equation (valid for all forms of the wave function)

$$J_r = \mathrm{Im}\left(\Psi^* \frac{\partial}{\partial r} \Psi\right) \quad . \tag{3.22}$$

The differential ionization cross section is equal to the ratio

$$d\sigma = \frac{dN}{k_o} \quad . \tag{3.23}$$

Using the asymptotic forms (2.20) and (2.23), we find

$$d\sigma = \frac{v_1 v_2}{k_o} |f(\underset{\sim}{v}_1,\underset{\sim}{v}_2)|^2 d\hat{\Omega}_1 d\hat{\Omega}_2 d\varepsilon_2 = \frac{\kappa}{k_o} |A(\hat{\Omega})|^2 d\hat{\Omega} \quad , \tag{3.24}$$

which agrees with the integrand in (3.11) and (3.15).

It is possible to measure dN experimentally as follows. Using two counters situated at distances r_1 and r_2 from the scattering center in the directions $\hat{\Omega}_1$ and $\hat{\Omega}_2$ and operating in coincidence (but with a given time delay T), we record the ionization events (per unit time), in which the first electron passes through the element of the sphere dS_1 and the second electron passes through the element of the sphere dS_2 within some increment of time dT, where

$$dS_1 = r_1^{\,2} d\hat{\Omega}_1 \quad , \quad dS_2 = r_2^{\,2} d\hat{\Omega}_2 \quad , \quad T = t_1 - t_2 \quad , \tag{3.25}$$

and t_1 and t_2 are the times the first and second electrons pass through the respective surface elements. The velocities of the departing electrons in this case are determined from the equations

$$r_1 = v_1 t_1 \quad , \quad r_2 = v_2 t_2 \quad . \tag{3.26}$$

Such a scheme was carried out by Cvejanović and Read [14], who made the measurements for $r_1 = r_2$ and different T. Their work will be discussed in §28. Here we investigate the case in which r_1 and r_2 are different, and the measurements are taken in the neighborhood of T = 0.

The number of ionization events recorded can be written in the form

$$dN = n(\underset{\sim}{r}_1, \underset{\sim}{r}_2, T) r_1^2 r_2^2 d\hat{\Omega}_1 d\hat{\Omega}_2 dT \quad . \tag{3.27}$$

From (3.26) we obtain

$$T = \frac{r_1}{v_1} - \frac{r_2}{v_2} \quad . \tag{3.28}$$

Differentiating (3.28) with constant r_1 and r_2, we find for T = 0,

$$dT = \frac{\kappa r}{v_1^2 v_2^2} d\varepsilon_2 \quad . \tag{3.29}$$

Equation (3.27) for T = 0 can be rewritten in the form

$$dN = r^5 \kappa^{-3} n(\underset{\sim}{r}_1, \underset{\sim}{r}_2, 0) d\hat{\Omega}_1 d\hat{\Omega}_2 d\varepsilon_2 \quad . \tag{3.30}$$

Comparing (3.30) with (3.20) and taking into account (3.14) and (3.16), we obtain

$$J_r(\underset{\sim}{r}_1, \underset{\sim}{r}_2) = \frac{\kappa}{v_1 v_2} n(\underset{\sim}{r}_1, \underset{\sim}{r}_2, 0) \quad . \tag{3.31}$$

Thus, by recording coincidences in the passage of the electrons through the surface elements dS_1 and dS_2 in the time interval dT, we are measuring the probability flux in configuration space.

Another way of experimentally determining the differential ionization cross section is to record the ionization events in which the electrons escape in certain directions with definite energies, as has been done by Ehrhardt *et al.* [15,16].

§4. *Flux Conservation in Ionization*

It follows from Eq. (3.19) that the probability flux through any closed hypersurface in the six-dimensional configuration space is zero.

We shall calculate the probability flux through finite sections of the following two infinite hypersurfaces. The first is defined by the condition

$$r_1 = \text{const} = R_1 \quad , \tag{4.1}$$

and the second by

$$r_2 = \text{const} = R_2 \quad . \tag{4.2}$$

The two hypersurfaces intersect at the points of the six-dimensional space where (4.1) and (4.2) are simultaneously satisfied. Let us consider the closed hypersurface composed of the section of the first hypersurface on which $r_2 \leq R_2$ and the section of the second on which $r_1 \leq R_1$. It is not difficult to see that every line in the six-dimensional space passing through the origin, i.e., determined by Eqs. (2.4), where $\underset{\sim}{v}_1$ and $\underset{\sim}{v}_2$ are constants and t is a parameter, intersects the chosen hypersurface at one and only one point.

The conservation of probability can be written in the form

$$X_1(R_1,R_2) + X_2(R_1,R_2) = 0 \quad , \tag{4.3}$$

where X_1 and X_2 are the fluxes through the first and second sections, respectively. The normals to the first and second hypersurfaces point in the directions in which r_1 and r_2 increase. Therefore, we obtain the fluxes in the directions of the normals by differentiating Ψ with respect to r_1 and r_2; thus,

$$X_1 = r_1^2 \int_{r_2 \le R_2} \mathrm{Im}\left(\Psi^* \frac{\partial}{\partial r_1} \Psi\right) d\underset{\sim}{r}_2 d\hat{\Omega}_1 \ , \ r_1 = R_1 \quad , \tag{4.4}$$

$$X_2 = r_2^2 \int_{r_1 \le R_1} \mathrm{Im}\left(\Psi^* \frac{\partial}{\partial r_2} \Psi\right) d\underset{\sim}{r}_1 d\hat{\Omega}_2 \ , \ r_2 = R_2 \quad . \tag{4.5}$$

Let us investigate the behavior of X_1 and X_2 when R_1 and $R_2 \to \infty$. It turns out that the result depends on the order in which we take the limits. Let us look first at the behavior of X_1 when R_1 is finite but large enough that the asymptotic form (1.3) is applicable, and $R_2 \to \infty$. It is not difficult to see that $X_1(R_1,\infty)$ is equal to the flux, determined by Eq. (3.5), of the first electron through a sphere of radius R_1 in three-dimensional space. We use the asymptotic expression for a plane wave given in §15:

$$e^{i\underset{\sim}{k}_o \cdot \underset{\sim}{r}_1} \sim \delta(\hat{\Omega}_o - \hat{\Omega}_1) \frac{2\pi}{ik_o r_1} e^{ik_o r_1} - \delta(\hat{\Omega}_o + \hat{\Omega}_1) \frac{2\pi}{ik_o r_1} e^{-ik_o r_1} \quad . \tag{4.6}$$

Then we obtain

$$\lim_{R_1 \to \infty} \lim_{R_2 \to \infty} X_1(R_1,R_2) = -4\pi \, \mathrm{Im}\, f_o(0) + X_1^d + X_1^{ion} \quad , \tag{4.7}$$

where

$$X_1^d = \sum_{n\ell m} k_n \int |f_{n\ell m}(\hat{\Omega}_1)|^2 d\hat{\Omega}_1 \quad , \tag{4.8}$$

$$X_1^{ion} = \int_{v\leq\kappa} k|f_{\underset{\sim}{v}}(\hat{\Omega}_1)|^2 d\underset{\sim}{v} d\hat{\Omega}_1 \quad . \tag{4.9}$$

If we change the order in which we take the limits, then the result for the part of the flux corresponding to the discrete spectrum does not change. For that part corresponding to ionization we get the integral

$$\int_{v,\tilde{v}\leq\kappa} Y(\underset{\sim}{v},\tilde{\underset{\sim}{v}}) f_{\underset{\sim}{v}}(\hat{\Omega}_1) f^*_{\tilde{\underset{\sim}{v}}}(\hat{\Omega}_1) k e^{i(k-\tilde{k})R_1 + i(\eta-\tilde{\eta})} d\underset{\sim}{v}\, d\tilde{\underset{\sim}{v}}\, d\hat{\Omega}_1 \quad , \tag{4.10}$$

where

$$Y(\underset{\sim}{v},\tilde{\underset{\sim}{v}}) = \int_{r_2\leq R_2} \phi(\underset{\sim}{v},\underset{\sim}{r}_2)\phi^*(\tilde{\underset{\sim}{v}},\underset{\sim}{r}_2) d\underset{\sim}{r}_2 \quad , \tag{4.11}$$

and the quantities $\tilde{k}$ and $\tilde{\eta}$ refer to $\tilde{\underset{\sim}{v}}$.

If R_2 is finite, then $Y(\underset{\sim}{v},\tilde{\underset{\sim}{v}}) \neq \delta(\underset{\sim}{v}-\tilde{\underset{\sim}{v}})$, and the integral (4.10) goes to zero as $R_1 \to \infty$ because the integrand oscillates infinitely fast. Hence,

$$\lim_{R_2\to\infty} \lim_{R_1\to\infty} X_1(R_1,R_2) = -4\pi \operatorname{Im} f_o(0) + X_1^d \quad . \tag{4.12}$$

This result has an obvious geometrical interpretation. Since the wave functions of the discrete spectrum fall off quickly as the argument increases, the probability flux corresponding to the excitation of discrete levels flows only in the specific directions of configuration space for which r_1 or r_2 remains finite. The ionization flux flows in all of the remaining directions of configuration space and forms a spherically diverging beam. Increasing R_2 for a given R_1 means increasing the dimensions of the first section of the hypersurface without getting farther away from the center (the origin of coordinates), and increasing R_1 for a given R_2 means removing the hypersurface from the center without any appreciable growth in the dimensions of the section. Hence, we find finally that in the first case fluxes in all directions pass through the hypersurface in question with the exception of those parallel to the hypersurface (corresponding to exchange excitation of discrete levels), and in the second case only those fluxes perpendicular to the hypersurface pass through it (corresponding to direct excitation of discrete levels). This can be seen more clearly in the example [17] in which the electrons are confined to one-dimensional motion.

Analogous to the derivation of (4.7) and (4.12), we obtain

$$\lim_{R_1\to\infty} \lim_{R_2\to\infty} X_2(R_1,R_2) = X_2^d \quad , \tag{4.13}$$

$$\lim_{R_2\to\infty} \lim_{R_1\to\infty} X_2(R_1,R_2) = X_2^d + X_2^{ion} \quad , \tag{4.14}$$

where

$$X_2^d = \sum_{n\ell m} k_n \int |g_{n\ell m}(\hat{\Omega}_2)|^2 d\hat{\Omega}_2 \quad , \tag{4.15}$$

$$X_2^{ion} = \int_{v\leq\kappa} k|g_{\underset{\sim}{v}}(\hat{\Omega}_2)|^2 d\underset{\sim}{v} d\hat{\Omega}_2 \quad . \tag{4.16}$$

Equation (2.3) implies

$$X_1^{ion} = X_2^{ion} \quad . \tag{4.17}$$

Equations (4.7) and (4.13) show that the limiting process, in which $R_2 \to \infty$ first, leads to the following expression for the conservation of flux:

$$X_1^d + X_2^d + X_1^{ion} = 4\pi \, \mathrm{Im} \, f_o(0) \quad . \tag{4.18}$$

If $R_1 \to \infty$ first, then from (4.12) and (4.14) we get an expression with X_1^{ion} replaced by X_2^{ion}, which because of (4.17) does not change the result. Dividing (4.18) by the incident-electron flux density, we get the optical theorem

$$\sigma_d + \sigma_{ion} = \frac{4\pi}{k_o} \, \mathrm{Im} \, f_o(0) \quad . \tag{4.19}$$

The fluxes of the first and second electrons through an infinitely distant sphere in three-dimensional space are determined from Eqs. (4.7) and (4.14). We see from (4.18) that they are nonzero, i.e., the individual electron fluxes in three-dimensional space are not conserved. According to (4.14)-(4.16) the flux of the second electron is positive. This is true because the second electron can only be ejected from the atom. By comparing (4.7) with (4.18) we obtain a negative quantity for the flux of the first electron

$$s_1 = -X_2^d \quad , \tag{4.20}$$

as in the exchange excitation of discrete levels the incident electron stays in the vicinity of the nucleus. When ionization occurs the total three-dimensional flux of both electrons is also not conserved. It follows from (4.7), (4.14), and (4.18) that

$$s_1 + s_2 = X_2^{ion} \quad . \tag{4.21}$$

This expression states that the total electron flux is increased by the electrons ejected from the atom during ionization.

§5. *Structure of the Wave Function*

For a more transparent representation of the various parts of the wave function contributing to the effective cross sections for direct and exchange excitation of discrete levels and for ionization of the atom, it is useful to look at the structure of the wave function when it is expanded in a series of atomic wave functions [18-20]. According to the results of §17, the integral (5.7) used below has a logarithmically diverging phase in the case of Coulomb interaction. Therefore, in the present section we shall consider only short-range potentials. We write the Hamiltonian of the two-electron system in the form

$$H = -\frac{1}{2}\Delta_1 - \frac{1}{2}\Delta_2 + V_o(r_1) + V_o(r_2) + V_{12}(r_{12}) \quad , \tag{5.1}$$

where $V_o(x)$ and $V_{12}(x)$ fall off sufficiently rapidly as $x \to \infty$.

The eigenfunctions of the atom satisfy the equation

$$\left[-\frac{1}{2}\Delta + V_o(r) - \varepsilon_\beta\right]\phi_\beta(\underset{\sim}{r}) = 0 \quad . \tag{5.2}$$

The index β (and also γ below) denotes the set of quantum numbers of both the discrete and the continuous spectra. We assume that in the continuous spectrum the function ϕ_β behaves asymptotically as a plane wave + diverging wave:

$$\phi_\beta(\underset{\sim}{r}) \equiv \phi_{\underset{\sim}{v}}(\underset{\sim}{r}) \sim (2\pi)^{-3/2}\left[e^{i\underset{\sim}{v}\cdot\underset{\sim}{r}} + \frac{t_v^*(-\hat{\Omega})}{r}\, e^{-ivr}\right] \quad , \tag{5.3}$$

where t_v is the scattering amplitude for the potential V_o.

We shall look for a solution to the Schrödinger equation

$$(H-E)\Psi = 0 \tag{5.4}$$

in the form of a series

$$\Psi = \sum_{\beta\gamma} C_{\beta\gamma}\phi_\beta(\underset{\sim}{r}_1)\phi_\gamma(\underset{\sim}{r}_2) \quad , \tag{5.5}$$

where the sums over β and γ also include integrals over the continuous spectrum. Substituting (5.5) into (5.4), we obtain the equation

$$(E-\varepsilon_\beta-\varepsilon_\gamma)C_{\beta\gamma} = a_{\beta\gamma} \quad , \tag{5.6}$$

where

$$a_{\beta\gamma} = \int \phi^*_\beta(\underset{\sim}{r}_1)\phi^*_\gamma(\underset{\sim}{r}_2)V_{12}(r_{12})\Psi(\underset{\sim}{r}_1,\underset{\sim}{r}_2)d\underset{\sim}{r}_1 d\underset{\sim}{r}_2 \quad . \tag{5.7}$$

The solution to the Schrödinger equation (5.4) that satisfies the boundary conditions (1.3) and (1.4) can be expressed in the form

$$\Psi = \Psi_o + \sum_{\beta\gamma} \frac{a_{\beta\gamma}\phi_\beta(\underset{\sim}{r}_1)\phi_\gamma(\underset{\sim}{r}_2)}{E-\varepsilon_\beta-\varepsilon_\gamma+i0} \quad , \tag{5.8}$$

where

$$\Psi_o = (2\pi)^{3/2}\phi^*_{-\underset{\sim}{k}_o}(\underset{\sim}{r}_1)\phi_o(\underset{\sim}{r}_2) \quad , \tag{5.9}$$

and $\underset{\sim}{k}_o$ is the wave vector of the incident electron.

The expansion (5.8) shows the structure of the wave function. The addition to E of an infinitesimally small positive imaginary part, which is equivalent to passing under the singularity in the complex ε plane, ensures the asymptotic behavior as a diverging wave. We note that the notation $\delta_+(x)$ is frequently used, which is defined by

$$\frac{1}{x+i0} = -2\pi i\delta_+(x) \quad . \tag{5.10}$$

The function δ_+ has the meaning of a δ function which isolates the diverging waves. If the integrand contains $\exp(ivr\alpha)$, where $v > 0$ and $r \to \infty$, then for $\alpha > 0$ the function $\delta_+(E - v^2/2)$ is equivalent to the usual δ function; for $\alpha < 0$ it gives a zero result.

The terms in expansion (5.8) satisfying the conservation of energy

$$\varepsilon_\beta + \varepsilon_\gamma = E \tag{5.11}$$

are singular. If at least one of ε_β and ε_γ satisfying (5.11) is in the continuous spectrum, then the singularity is removed by the indicated choice of integration path. If, however, they are both in the discrete spectrum, the singularity is not removable. In this case we should expect the corresponding coefficient $a_{\beta\gamma}$ to go to zero. This case cannot occur in the ionization problem when $E > 0$.

Only the singular terms in the sum (5.8) correspond to real collision processes, and contribute directly to the effective cross sections. In this case ε_β and ε_γ are the energies of the first and second electrons after the collision, and the coefficient $a_{\beta\gamma}$ is proportional to the scattering amplitude. The remaining terms in (5.8) refer to virtual processes.

In order to demonstrate this, we consider (5.8) for $r_1 \to \infty$. Then, in the sum over β we may ignore the rapidly decaying terms from the discrete spectrum. In the continuous spectrum $\beta \equiv \underset{\sim}{v}$. Using Eqs. (5.3), (4.6), and (5.10), we obtain for the integral over the continuous spectrum

$$\int \frac{a_{\underset{\sim}{v}\gamma}\phi_{\underset{\sim}{v}}(\underset{\sim}{r}_1)d\underset{\sim}{v}}{E - v^2/2 - \varepsilon_\gamma + i0} \sim -\sqrt{2\pi}\, a_{\underset{\sim}{k}\gamma}\, r_1^{-1}\, e^{ik_\gamma r_1} \quad , \tag{5.12}$$

where

$$k_\gamma = \sqrt{2(E-\varepsilon_\gamma)} \quad , \quad \underset{\sim}{k} = \frac{k_\gamma}{r_1}\underset{\sim}{r}_1 \quad . \tag{5.13}$$

This implies that

$$\Psi \sim \Psi_o - \sqrt{2\pi}\, r_1^{-1} \sum_{\varepsilon_\gamma \le E} \phi_\gamma(\underset{\sim}{r}_2) a_{\underset{\sim}{k}\gamma} e^{ik_\gamma r_1} \quad . \tag{5.14}$$

Similarly, for $r_2 \to \infty$ we obtain

$$\Psi \sim -\sqrt{2\pi}\, r_2^{-1} \sum_{\varepsilon_\beta \le E} \phi_\beta(\underset{\sim}{r}_1) a_{\beta\underset{\sim}{k}} e^{ik_\beta r_2} \quad , \tag{5.15}$$

where

$$k_\beta = \sqrt{2(E-\varepsilon_\beta)} \quad , \quad \underset{\sim}{k} = \frac{k_\beta}{r_2}\underset{\sim}{r}_2 \quad . \tag{5.16}$$

Comparison of these equations with (1.3) and (1.4), where for short-range forces we assume $\eta = 0$, shows that

$$-\sqrt{2\pi}\, a_{\underset{\sim}{k}\underset{\sim}{v}} = f(\underset{\sim}{k},\underset{\sim}{v}) = g(\underset{\sim}{v},\underset{\sim}{k}) \quad . \tag{5.17}$$

This confirms Eq. (2.3). The relation (5.17) demonstrates that the same terms of expansion (5.8) represent direct and exchange ionization.

Figure 1 shows the singular terms of expansion (5.8) in the $\varepsilon_\beta\varepsilon_\gamma$ plane. The singular points are situated along the line FG defined by Eq. (5.11). They are divided into three

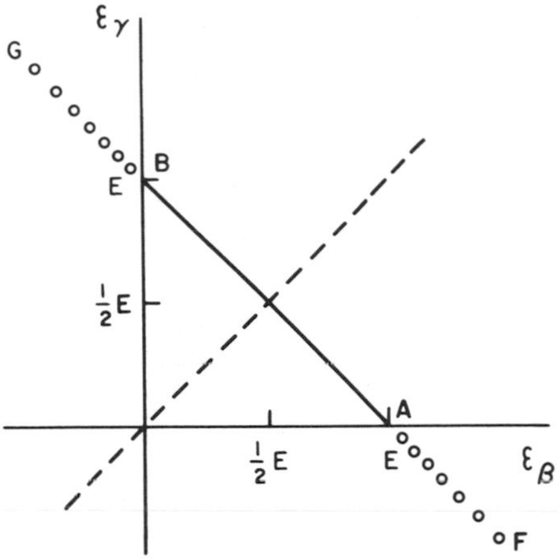

Fig. 1. Singular points in the plane of the two electron energies.

groups. The points of FA are defined by the conditions

$$\varepsilon_\beta > 0 \ , \ \varepsilon_\gamma < 0 \tag{5.18}$$

and correspond to direct excitation of discrete levels. The points of GB, corresponding to exchange excitation, satisfy

$$\varepsilon_\beta < 0 \ , \ \varepsilon_\gamma > 0 \quad . \tag{5.19}$$

For the third group of points we have

$$\varepsilon_\beta > 0 \ , \ \varepsilon_\gamma > 0 \quad . \tag{5.20}$$

These points form the solid line AB and correspond to ionization. In this group the division into direct and exchange processes follows convention. It seems natural to attribute those cases with $\varepsilon_\beta > \varepsilon_\gamma$ to direct processes and those with $\varepsilon_\beta < \varepsilon_\gamma$ to exchange processes.

In practical calculations an expansion in a single series of atomic wave functions is often used. The simplest such expansions can be written in the form

$$\Psi = \sum_{\beta} X_{\beta}(\underset{\sim}{r}_1)\phi_{\beta}(\underset{\sim}{r}_2) \tag{5.21}$$

or

$$\Psi = \sum_{\beta} Y_{\beta}(\underset{\sim}{r}_2)\phi_{\beta}(\underset{\sim}{r}_1) \quad . \tag{5.22}$$

Comparing these with (5.8), we find

$$X_{\beta} = X_{\beta o} + \sum_{\gamma} \frac{a_{\gamma\beta}\phi_{\gamma}}{E-\varepsilon_{\beta}-\varepsilon_{\gamma}+i0} \quad , \tag{5.23}$$

$$Y_{\beta} = Y_{\beta o} + \sum_{\gamma} \frac{a_{\beta\gamma}\phi_{\gamma}}{E-\varepsilon_{\beta}-\varepsilon_{\gamma}+i0} \quad , \tag{5.24}$$

where $X_{\beta o}$ and $Y_{\beta o}$ are the contributions to X_{β} and Y_{β} from the function Ψ_o, defined by (5.9).

The series (5.23) and (5.24) may contain a singular term in which $\varepsilon_{\gamma} = E - \varepsilon_{\beta}$. If $E > \varepsilon_{\beta}$, then $\varepsilon_{\gamma} > 0$, and the singularity is removed by the choice of integration path. If, however, $E - \varepsilon_{\beta} < 0$ and E is equal to the energy ε_{γ} of some discrete level, and moreover $\varepsilon_{\beta} > 0$, then X_{β} and Y_{β} contain a nonremovable singularity. The singularities of X_{β} correspond to exchange excitation of discrete levels, and those of Y_{β} to direct excitation. Thus, (5.21) describes direct processes well and exchange processes poorly, while for (5.22) the opposite holds.

In order to avoid the singularities, we should use an expansion in which the direct and exchange processes are each described by appropriate groups of terms:

$$\Psi = \sum_{\beta} [F_{\beta}(\underset{\sim}{r}_1)\phi_{\beta}(\underset{\sim}{r}_2) + G_{\beta}(\underset{\sim}{r}_2)\phi_{\beta}(\underset{\sim}{r}_1)] \quad . \tag{5.25}$$

As distinct from the expansions (5.21) and (5.22), in which X_{β} and Y_{β} are uniquely defined, F_{β} and G_{β} in (5.25) are not unambiguously defined. It is easy to convince oneself that Ψ does not change under the following transformation:

$$F_\beta \to F_\beta + \sum_\gamma c_{\beta\gamma}\phi_\gamma \quad , \quad G_\beta \to G_\beta - \sum_\gamma c_{\gamma\beta}\phi_\gamma \quad , \tag{5.26}$$

where $c_{\beta\gamma}$ are arbitrary coefficients.

In order that F_β and G_β be nonsingular, i.e., have no singularities in the discrete spectrum, it is necessary that the first part of the sum (5.25) contain the singular terms FA (see Fig. 1) and the second part contain the terms GB. But this condition fails to define F_β and G_β uniquely. We can require that the first part of (5.25) contain all singular and nonsingular terms of the sum (5.8) with $\varepsilon_\beta \geq \varepsilon_\gamma$, and the second part contain the terms with $\varepsilon_\beta \leq \varepsilon_\gamma$. This means dividing the $\varepsilon_\beta\varepsilon_\gamma$ plane along the bisector $\varepsilon_\beta = \varepsilon_\gamma$. Then, we have

$$F_\beta = F_{\beta o} + \sum_{\substack{\gamma \\ \varepsilon_\gamma \geq \varepsilon_\beta}} \frac{a_{\gamma\beta}\phi_\gamma}{E - \varepsilon_\beta - \varepsilon_\gamma + i0} \quad , \tag{5.27}$$

$$G_\beta = \sum_{\substack{\gamma \\ \varepsilon_\gamma \geq \varepsilon_\beta}} \frac{a_{\beta\gamma}\phi_\gamma}{E - \varepsilon_\beta - \varepsilon_\gamma + i0} \quad . \tag{5.28}$$

Note that (5.27) differs from (5.8) by a transposition of the indices β and γ.

It follows from (5.27) and (5.28) that F_β and G_β do not contain any atomic states with lower energy:

$$\int F_\beta \phi_\gamma^* d\underset{\sim}{r} = \int G_\beta \phi_\gamma^* d\underset{\sim}{r} = 0 \quad , \quad \varepsilon_\gamma < \varepsilon_\beta \quad . \tag{5.29}$$

This property can be used to determine F_β and G_β in practical calculations.

It is not difficult to see that if $\varepsilon_\beta > E/2$, then (5.27) and (5.28) for the continuous spectrum also cannot contain any singular terms, and consequently do not contribute to the effective cross sections. Taking this fact into account, we have the following asymptotic expressions for $r \to \infty$:

$$F_\beta \sim \delta_{\beta o} e^{i\mathbf{k}_o \cdot \mathbf{r}} + r^{-1} f_\beta(\Omega) e^{ik_\beta r} \quad , \tag{5.30}$$

$$G_\beta \sim r^{-1} g_\beta(\Omega) e^{ik_\beta r} \quad , \quad \varepsilon_\beta \leq \frac{1}{2} E \quad , \tag{5.31}$$

$$F_\beta = o(r^{-1}) \ , \ G_\beta = o(r^{-1}) \ , \ \varepsilon_\beta > \frac{1}{2} E \quad . \tag{5.32}$$

The asymptotic forms (5.30)-(5.32) agree with Eq. (3.13), in which the integration over energy is limited to $0 \leq \varepsilon \leq E/2$. They differ from Eqs. (5.14) and (5.15), in which the summation covers the range $\varepsilon \leq E$, because the functions F_β and G_β contribute to both (5.14) and (5.15).

CHAPTER II

ASYMPTOTIC EXPANSION OF THE WAVE FUNCTION FOR A SYSTEM OF CHARGED PARTICLES

§6. *Asymptotic Expansion of the Wave Function for the Ionization of Hydrogen by an Electron*

The characteristic nature of the slowly decaying Coulomb interaction of charged particles causes the wave function at large distances to differ from the wave function for free motion by a logarithmically increasing term in the phase. The asymptotic behavior of the wave function for the case of two charged particles is known. In the problem of the ionization of an atom by an electron the system in the final state contains three or more charged particles, moving with positive energies. It is to be expected that in this case as well the presence of the Coulomb interaction leads to a logarithmic phase shift. In accordance with this expectation, in the boundary conditions (1.3) and (1.4) an additional term η was introduced into the phase of the scattered wave.

For the ionization problem the asymptotic domain where both electrons are found far from the nucleus is of interest. In this domain the asymptotic forms (1.3) and (1.4) can be reduced, by means of the method of stationary phase, to (2.20), (2.22), and (2.23), which describe a spherically diverging wave in six-dimensional configuration space. Later we shall investigate the asymptotic behavior in the form of (2.23). The logarithmic term in the phase can be determined by two methods: by constructing a formal solution of the Schrödinger equation in the form of an asymptotic series [21], and by using the semiclassical approximation [21,22]. In this chapter we shall study the solution to the Schrödinger equation in the form of an asymptotic series. First let us look at the asymptotic solution for the ionization of the hydrogen atom by an electron.

To simplify the exposition, we investigate the case in which the total orbital angular momentum of the electrons is $L = 0$. Then, the wave function depends only on three variables, r_1, r_2, and θ, the angle between $\underset{\sim}{r}_1$ and $\underset{\sim}{r}_2$. We introduce the hyperspherical coordinates r and α, defined in (2.6) and (2.7). The potential energy is written in the form

$$V = -\frac{Z(\hat{\Omega})}{r} \quad . \tag{6.1}$$

Here $\hat{\Omega}$ denotes the quantities α and θ.

From (1.2) we can rewrite (6.1) as

$$Z(\hat{\Omega}) = \frac{1}{\cos\alpha} + \frac{1}{\sin\alpha} - \frac{1}{\sqrt{1-\cos\theta\sin 2\alpha}} \quad . \tag{6.2}$$

Equation (1.1) takes the form

$$\left[\frac{1}{r^5}\cdot\frac{\partial}{\partial r}\left(r^5\frac{\partial}{\partial r}\right) + \frac{\Delta^*}{r^2} + \frac{2Z(\hat{\Omega})}{r} + 2E\right]\Psi = 0 \quad , \tag{6.3}$$

where

$$\Delta^* = \frac{1}{\sin^2 2\alpha}\left[\frac{\partial}{\partial\alpha}\left(\sin^2 2\alpha\,\frac{\partial}{\partial\alpha}\right) + \frac{4}{\sin\theta}\cdot\frac{\partial}{\partial\theta}\left(\sin\theta\,\frac{\partial}{\partial\theta}\right)\right] . \tag{6.4}$$

We shall look for a solution to Eq. (6.3) in the form of a series

$$\Psi = r^{-5/2}e^{i\kappa r + iW(\hat{\Omega})\,\ln\kappa r}\sum_{nm} A_{nm}(\hat{\Omega})\,\frac{(\ln\kappa r)^m}{(\kappa r)^n} \quad , \tag{6.5}$$

where the quantities κ, $W(\hat{\Omega})$, and $A_{nm}(\hat{\Omega})$ are to be specified. The expansion in powers of κr is performed to simplify the later equations. The form of the series is based on the following considerations: 1) the first term of the series should be analogous to (2.23); 2) it is well known that the solution to Eq. (1.1) for small r cannot be expanded in powers of r, but exists in the form of a double series in powers of r and ℓn r [23-25]. Later it will be shown that it is also necessary in (6.5) to retain powers of the logarithms.

The expansion in the double series in powers of r and ℓn r is not unique, since a power of r can be expressed as an infinite sum of powers of ℓn r. Therefore, we should require that for each n the series in m be finite or converge sufficiently rapidly.

Substitution of (6.5) into (6.3) leads to the recursion relation

$$2\left(n - iW + \frac{iZ}{\kappa}\right) A_{nm} = i\left(1 - \frac{2E}{\kappa^2}\right) A_{n+1,m} + 2(m+1)A_{n,m+1} +$$
$$+ iUA_{n-1,m-2} + D_1 A_{n-1,m-1} - \{i[D_2 + n(n-1)] + (2n-1)W\}A_{n-1,m} +$$
$$+ (m+1)[2W + i(2n-1)]A_{n-1,m+1} - i(m+1)(m+2)A_{n-1,m+2} , \qquad (6.6)$$

where U is the function

$$U = \left(\frac{\partial W}{\partial \alpha}\right)^2 + \frac{4}{\sin^2 2\alpha}\left(\frac{\partial W}{\partial \theta}\right)^2 , \qquad (6.7)$$

and D_1 and D_2 are differential operators

$$D_1 = 2\,\frac{\partial W}{\partial \alpha}\cdot\frac{\partial}{\partial \alpha} + \frac{8}{\sin^2 2\alpha}\cdot\frac{\partial W}{\partial \theta}\cdot\frac{\partial}{\partial \theta} + (\Delta^* W) , \qquad (6.8)$$

$$D_2 = \Delta^* - \frac{15}{4} - W^2 . \qquad (6.9)$$

In order that the series (6.5) correspond to the asymptotic form (2.23), it should not contain any positive powers of r, and for n = 0 it should not contain any positive powers of ℓn r, i.e., we require that

$$A_{nm} = 0 \quad , \quad n < 0 \quad , \tag{6.10}$$

$$A_{0m} = 0 \quad , \quad m > 0 \quad . \tag{6.11}$$

It is assumed that A_{00} is not identically equal to zero. Then, when (6.10) is taken into account, the recursion relation (6.6) for n = -1, m = 0 gives

$$\kappa^2 = 2E \quad , \tag{6.12}$$

which agrees with the value of κ in (2.23).

For n = m = 0, allowing for (6.11), we obtain

$$W(\hat{\Omega}) = \frac{Z(\hat{\Omega})}{\kappa} \quad . \tag{6.13}$$

Considering the recursion relation for n = 0, m = -2, -3,..., we obtain

$$A_{0m} = 0 \quad , \quad m < 0 \quad . \tag{6.14}$$

The condition (6.11) is equivalent to (6.13) and states that for n = 0 the implied sum over powers of the logarithms is equivalent to the logarithmic term in the phase of the wave function.

We shall consider (6.10) and (6.13) to be the principal conditions. They are not arbitrary, but rather are the only ones allowed by the recursion relation. To show this, suppose that W has some arbitrary value, and the series (6.5) begins with some arbitrary $n = n_o$. Then, instead of (6.10) we have

$$A_{nm} = 0 \quad , \quad n < n_o \quad . \tag{6.15}$$

For $n = n_o - 1$, from the recursion relation we again obtain (6.12), and for $n = n_o$ we get

$$(n_o - iW + iZ/\kappa)A_{n_o m} = (m+1)A_{n_o,m+1} \quad . \qquad (6.16)$$

Considering (6.16) for $m = -1,-2,\ldots$, we find

$$A_{n_o m} = 0 \quad , \quad m < 0 \quad . \qquad (6.17)$$

For $m > 0$ we obtain

$$A_{n_o m} = \frac{[n_o - iW + (iZ/\kappa)]^m}{m!} A_{n_o o} \quad . \qquad (6.18)$$

This implies

$$e^{i\rho+iW\,\ell n\,\rho}\,\rho^{-n_o}\sum_m A_{n_o m}(\ell n\ \rho)^m = A_{n_o o}e^{i\rho+(iZ/\kappa)\ell n\,\rho}, \qquad (6.19)$$

where we have introduced the notation

$$\rho = \kappa r \quad . \qquad (6.20)$$

Equation (6.19) shows that for an arbitrary choice of n_o and W the sum of the leading terms will correspond to (6.10)-(6.14). For $n_o \neq 0$ the series in m turns out to be infinite and its sum compensates for the nonzero value of n_o. It is only when $n_o = 0$ and condition (6.13) holds that the series in powers of the logarithms is finite. The choice of other values for n_o and W means a rearrangement of the same series into a considerably more complicated form. Henceforth we shall assume that $n_o = 0$ and that W is given by (6.13).

Let us consider now the recursion relation for $n = 1$. Substituting $m = -1,-2,\ldots$ into (6.6) and taking into account (6.14), we obtain

$$A_{1m} = 0 \quad , \quad m < 0 \quad . \tag{6.21}$$

Substituting m = 0,1,2,..., we obtain

$$2A_{10} = 2A_{11} - iD_2A_{00} \quad , \tag{6.22}$$

$$2A_{11} = 4A_{12} + D_1A_{00} \quad , \tag{6.23}$$

$$2A_{12} = 6A_{13} + iUA_{00} \quad , \tag{6.24}$$

$$A_{1m} = (m+1)A_{1,m+1} \quad , \quad m \geq 3 \quad . \tag{6.25}$$

From (6.25) we find

$$A_{1m} = \frac{3!A_{13}}{m!} \quad , \quad m \geq 3 \quad . \tag{6.26}$$

We may choose A_{13} arbitrarily. Then, from (6.22)-(6.24) we can determine A_{12}, A_{11}, and A_{10}. As in the above discussion of the choice of n_o, we can show that one should set $A_{13} = 0$. Indeed, from (6.26) it follows that

$$\frac{1}{\rho} \sum_{m=3}^{\infty} A_{1m}(\ln \rho)^m = 3!A_{13} \left[1 - \frac{1}{\rho} - \frac{\ln \rho}{\rho} - \frac{(\ln \rho)^2}{2\rho}\right] \quad . \tag{6.27}$$

We see that the sum over m ≥ 3 can be expressed in terms of the preceding terms of the expansion containing the coefficients A_{00}, A_{10}, A_{11}, and A_{12}. When (6.27) is added to the preceding terms in (6.5), one may set $A_{13} = 0$ in the resulting expression. Thus the expansion (6.5) is reduced to its simplest form.

If $A_{13} = 0$, we obtain from (6.22)-(6.24)

$$A_{10} = \left(iU + \frac{1}{2} D_1 - \frac{i}{2} D_2\right)A_{00} \quad , \tag{6.28}$$

$$A_{11} = \left(iU + \frac{1}{2} D_1\right)A_{00} \quad , \tag{6.29}$$

$$A_{12} = \frac{i}{2} UA_{00} \quad . \tag{6.30}$$

It is seen that the coefficients A_{1m} are expressed in terms of A_{00} and its derivatives.

We obtain the analogous results for n = 2,3,... by mathematical induction. If

$$A_{n-1,m} = 0 \quad , \quad m < 0 \quad , \tag{6.31}$$

then, considering the recursion relation (6.6) for m = -1, -2,..., it is not difficult to see also that

$$A_{nm} = 0 \quad , \quad m < 0 \quad . \tag{6.32}$$

If

$$A_{n-1,m} = 0 \quad , \quad m > 2(n-1) \quad , \tag{6.33}$$

then from the recursion relation we obtain

$$A_{nm} = \frac{(2n+1)!}{n^{2n+1}} \cdot \frac{n^m}{m!} A_{n,2n+1} \quad , \quad m \geq 2n+1 \quad . \tag{6.34}$$

This implies

$$\frac{1}{\rho^n} \sum_{m=2n+1}^{\infty} A_{nm}(\ln \rho)^m = \frac{(2n+1)!}{n^{2n+3}} A_{n,2n+1}\left[1 - \frac{1}{\rho^n} \sum_{m=0}^{2n} \frac{n^m(\ln \rho)^m}{m!}\right] , \tag{6.35}$$

i.e., the sum over m > 2n is expressed in terms of the preceding terms. Therefore, we should set

$$A_{nm} = 0 \quad , \quad m > 2n \quad . \tag{6.36}$$

Then, using the recursion relation, we can successively find $A_{n,2n}, A_{n,2n-1}, \ldots, A_{no}$, if the coefficients $A_{n-1,m}$ $(m = 0,1,\ldots,2n-2)$ arc known, the new coefficients being expressed in terms of the previous ones and their derivatives. In the end all coefficients can be expressed in terms of A_{00} and W and their derivatives of various orders. In this manner the series (6.5) is determined uniquely, once A_{00} (α,θ) is given. The order of the derivatives increases with increasing index n and decreases with increasing index m. The coefficient A_{no} contains derivatives of A_{00} and W of order 2n with respect to α and θ.

The sum over m for given n in (6.5) is finite. But the sum over n for fixed m diverges, as the right side of the recursion relation (6.6) contains n^2. For large n we obtain

$$A_{nm} \approx -\frac{in}{2} A_{n-1,m} \quad , \qquad (6.37)$$

from which it follows that A_{nm} grows as n!. Hence, the series over n has meaning only in the asymptotic sense.

The presence of logarithmic terms is due to the dependence of W on α and θ. If W goes to a constant (i.e., $Z(\hat{\Omega}) \to$ const), then $U \to 0$, $D_1 \to 0$, and all coefficients A_{nm} for $m > 0$ vanish.

The expansion (6.5) contains powers of $\ell n\ r$, as does the Fock series [23-25], which is valid for small r and has the form

$$\Psi = \sum_{n=0}^{\infty} r^n \sum_{m=0}^{[n/2]} (\ell n\ r)^m a_{nm}(\hat{\Omega}) \quad . \qquad (6.38)$$

Here [n/2] denotes the integral part of n/2.

We note that (6.5) and (6.38) differ in the limits of the summation over m. Another difference is that in the Fock series subsequent coefficients are determined from the preceding ones by integration, while in (6.5) they are determined by differentiation.

The first term of the series (6.5) is of the form

$$\Psi \sim A_{00}(\hat{\Omega}) r^{-5/2} e^{i\kappa r + i[Z(\hat{\Omega})/\kappa] \ln \kappa r} \quad . \tag{6.39}$$

The expression (6.39) differs from the free-particle asymptotic form by the logarithmic shift in the phase. From the derivation of Eq. (6.19) it can be seen that the asymptotic form (6.39) is determined uniquely, and is independent of the initial choice of the function W in (6.5). In particular, we could have set W = 0 at the beginning.

The logarithmic phase shift in (6.39) is analogous in form to the phase shift in the well-known case of scattering of a single particle in a Coulomb field. The difference between the two cases is that for the usual Coulomb scattering Z = const, whereas in (6.39) Z depends on the direction $\hat{\Omega}$. For the scattering of three particles the logarithmic term in the phase also has a comparatively simple form with a physically clear interpretation. In the process of scattering the motion of the system is such that for $t \to \infty$ (t denotes time) its direction in configuration space approaches some constant value. The phase shift is determined by the magnitude of Z corresponding to this value of $\hat{\Omega}$. The other directions influence only the subsequent terms of the expansion.

Knowing the asymptotic form (6.39), we can determine the magnitude of η (see §21).

§7. *The Asymptotic Expansion in the General Case*

It is also not difficult to construct an asymptotic expansion in the general case of an arbitrary number of charged particles. To do this it is convenient to use hyperspherical coordinates. Therefore, we first present some information about this coordinate system [26].

If $x_1,\ldots,x_n$ are Cartesian coordinates in an n-dimensional space, then the hyperspherical coordinates $r,\theta_1,\ldots,\theta_{n-1}$ are defined by the equations

$$x_i = r\beta_i \cos\theta_i \quad , \quad i = 1,\ldots,n-1 \quad , \tag{7.1}$$

$$x_n = r\beta_{n-1} \sin\theta_{n-1} \quad , \tag{7.2}$$

where

$$\beta_1 = 1 \quad , \tag{7.3}$$

$$\beta_i = \sin\theta_1 \sin\theta_2 \ldots \sin\theta_{i-1} \, , \; i = 2,\ldots,n-1 \quad . \tag{7.4}$$

Equations (7.1)-(7.4) imply that

$$r^2 = x_1^2 + \ldots + x_n^2 \quad . \tag{7.5}$$

The angular variables are restricted to the intervals

$$0 \le \theta_i \le \pi \quad , \quad i = 1,\ldots,n-2 \quad , \tag{7.6}$$

$$0 \le \theta_{n-1} \le 2\pi \quad . \tag{7.7}$$

The n-dimensional volume element is of the form

$$d\underset{\sim}{r} = r^{n-1} d\hat{\Omega} dr \quad , \tag{7.8}$$

where the surface element on the unit sphere is defined by

$$d\hat{\Omega} = (\sin\theta_1)^{n-2}(\sin\theta_2)^{n-3} \ldots \sin\theta_{n-2} d\theta_1 \ldots d\theta_{n-1} \quad . \tag{7.9}$$

The surface area of the unit sphere is

$$\omega_n = 2\pi^{n/2}/\Gamma(n/2) \quad , \tag{7.10}$$

where Γ denotes the gamma function.

The line element is of the form

$$ds^2 = dr^2 + r^2 \sum_{i=1}^{n-1} \beta_i^2 d\theta_i^2 \quad . \tag{7.11}$$

The n-dimensional Laplacian in hyperspherical coordinates takes the form

$$\Delta = \Delta^{(r)} + r^{-2}\Delta^* \quad , \tag{7.12}$$

where

$$\Delta^{(r)} = \frac{1}{r^{n-1}} \cdot \frac{\partial}{\partial r}\left(r^{n-1}\frac{\partial}{\partial r}\right) \quad , \tag{7.13}$$

$$\Delta^* = \sum_{i=1}^{n-1} \frac{1}{\beta_i^2 (\sin\theta_i)^{n-1-i}} \cdot \frac{\partial}{\partial\theta_i}\left[(\sin\theta_i)^{n-1-i}\frac{\partial}{\partial\theta_i}\right] \quad . \tag{7.14}$$

The eigenvalues of the operator Δ^* are

$$-K(K+n-2) \quad , \quad K = 0,1,2,\ldots \quad . \tag{7.15}$$

The eigenfunctions of the operator Δ^* are called hyperspherical harmonics or K harmonics.

We consider a system of N+1 charged particles. Let $\underset{\sim}{r}_i$ be the radius vector, m_i the mass, and ζ_i the charge of the *ith* particle. The Schrödinger equation has the form

$$(T + V - E)\Psi = 0 \quad , \tag{7.16}$$

where

$$T = -\frac{\hbar^2}{2}\sum_{i=1}^{N+1}\frac{\Delta_i}{m_i} \quad , \tag{7.17}$$

$$V = \sum_{j<i} \frac{\zeta_i\zeta_j}{|\underset{\sim}{r}_i - \underset{\sim}{r}_j|} \quad . \tag{7.18}$$

We introduce a coordinate system in which the center-of-mass motion is separated out and in which the kinetic energy operator takes a homogeneous form. We may choose as such a system the set of normalized vectors $\underset{\sim}{q}_i$, which connect the (i+1)*th* particle with the center of mass of the first i particles, and radius vector $\underset{\approx}{R}$ to the center of mass of all the particles. The vectors $\underset{\sim}{q}_i$ are defined by [27]

$$\underset{\sim}{q}_i = \sqrt{\frac{m_{i+1}M_i}{M_{i+1}}}\,(\underset{\sim}{r}_{i+1} - \underset{\sim}{R}_i) \ , \ i = 1,\ldots,N \quad , \qquad (7.19)$$

where

$$M_i = m_1 + \ldots + m_i \quad , \qquad (7.20)$$

$$\underset{\sim}{R}_i = \frac{m_1\underset{\sim}{r}_1 + \ldots + m_i\underset{\sim}{r}_i}{M_i} \quad . \qquad (7.21)$$

The kinetic energy operator in the new coordinates takes the form

$$T = -\frac{\hbar^2}{M}\Delta_R - \hbar^2 \sum_{i=1}^{N} \Delta_{q_i} \quad , \qquad (7.22)$$

where M is the mass and $\underset{\approx}{R}$ is the radius vector of the center of mass of the total system. A similar relation holds also for the classical expression of the kinetic energy

$$T_{c\ell} = \frac{1}{2}\sum_{i=1}^{N+1} m_i \left(\frac{d\underset{\sim}{r}_i}{dt}\right)^2$$

$$= \frac{1}{2} M \left(\frac{d\underset{\approx}{R}}{dt}\right)^2 + \frac{1}{2}\sum_{i=1}^{N}\left(\frac{d\underset{\sim}{q}_i}{dt}\right)^2 \quad . \qquad (7.23)$$

Owing to the linearity of the transformation (7.19), Eq. (7.23) implies

$$\sum_{i=1}^{N+1} m_i r_i^2 = MR^2 + \sum_{i=1}^{N} q_i^2 \quad . \tag{7.24}$$

Henceforth we shall work with the Schrödinger equation for the relative motion

$$\left(\hbar^2 \sum_{i=1}^{N} \Delta_{q_i} - 2V + 2E\right)\Psi = 0 \quad . \tag{7.25}$$

Hereafter E denotes the energy in the center-of-mass system.

From (7.19) we obtain the expression for the vector $\underset{\sim}{r}_i$ in terms of $\underset{\sim}{q}_1,\ldots,\underset{\sim}{q}_N$ and $\underset{\sim}{R}$:

$$\underset{\sim}{r}_i = \underset{\sim}{R} + \sqrt{\frac{M_{i-1}}{m_i M_i}}\, \underset{\sim}{q}_{i-1} - \sum_{p=i}^{N} \sqrt{\frac{m_{p+1}}{M_p M_{p+1}}}\, \underset{\sim}{q}_p \quad . \tag{7.26}$$

When i=1 there is no term containing $\underset{\sim}{q}_{i-1}$, and when i = N+1 there is no sum over p.

From (7.26) we see that the difference $\underset{\sim}{r}_i-\underset{\sim}{r}_j$ is a linear homogeneous function of the vectors $\underset{\sim}{q}_1,\ldots,\underset{\sim}{q}_N$. Hence, if in the 3N-dimensional space of the vectors $\underset{\sim}{q}_1,\ldots,\underset{\sim}{q}_N$ we introduce hyperspherical coordinates, which again we denote by r and $\theta_1,\ldots,\theta_{3N-1}$, then the potential energy V can again be written in the form (6.1), where $\hat{\Omega}$ now denotes the set of coordinates $\theta_1,\ldots,\theta_{3N-1}$. Note that in the general case $Z(\hat{\Omega})$ has a considerably more complicated form than (6.2).

The Schrödinger equation (7.25) can be rewritten in a form analogous to (6.3):

$$\left[\frac{1}{r^{3N-1}} \cdot \frac{\partial}{\partial r}\left(r^{3N-1}\frac{\partial}{\partial r}\right) + \frac{\Delta^*}{r^2} + \frac{2\kappa W(\hat{\Omega})}{r} + \kappa^2\right]\Psi = 0 \quad . \tag{7.27}$$

Here Δ^* is defined by Eq. (7.14) for n = 3N, and

$$\kappa = \frac{\sqrt{2E}}{\hbar} \quad , \quad W(\hat{\Omega}) = \frac{Z(\hat{\Omega})}{\hbar\sqrt{2E}} \quad . \tag{7.28}$$

If we use a coordinate system in which R=0, then from (7.24) we obtain

$$r^2 \equiv \sum_{i=1}^{N} q_i^{\,2} = \sum_{i=1}^{N+1} m_i r_i^{\,2} \quad . \tag{7.29}$$

Substituting into (7.27) the expansion

$$\Psi = r^{-(3N-1/2)} e^{i\kappa r + iW(\hat{\Omega})\,\ell n\,\kappa r} \sum_{n\geq 0} (\kappa r)^{-n} \sum_{m=0}^{2n} (\ell n\;\kappa r)^m A_{nm}(\hat{\Omega}) \quad , \tag{7.30}$$

we once more find a recursion relation similar to (6.6). It differs from (6.6), however, in the following ways. On the left side and in the first term on the right, there are no terms containing W, Z, and $A_{n+1,m}$, since we have already used (7.28). Furthermore, instead of (6.7)-(6.9) we have

$$U = \sum_{i=1}^{3N-1} \frac{1}{\beta_i^{\,2}} \left(\frac{\partial W}{\partial \theta_i}\right)^2 \quad , \tag{7.31}$$

$$D_1 = 2 \sum_{i=1}^{3N-1} \frac{1}{\beta_i^{\,2}} \cdot \frac{\partial W}{\partial \theta_i} \cdot \frac{\partial}{\partial \theta_i} + (\Delta^* W) \quad , \tag{7.32}$$

$$D_2 = \Delta^* - \frac{(3N-1)(3N-3)}{4} - W^2 \quad . \tag{7.33}$$

The operator Δ^* is now, of course, defined by (7.14). Hence it is clear that the structure of the expansion [e.g., the Eqs. (6.28)-(6.30)] is not changed.

We note that (7.31) can be written in the form

$$U = r^2 (\nabla W)^2 \quad , \tag{7.34}$$

where ∇ denotes the gradient operator in 3N-dimensional space.

The asymptotic expression (6.5) retains its form when, in addition to the Coulomb potential, there is a more rapidly decaying potential which can be represented as a series in inverse powers of r:

$$V = -\frac{Z(\hat{\Omega})}{r} + \hbar^{-2} \sum_{j\geq 2} c_j(\hat{\Omega})\rho^{-j} \quad . \tag{7.35}$$

In this case the right side of the recursion relation contains the additional sum

$$-i \sum_{j\geq 1} c_{j+1} A_{n-j,m} \quad . \tag{7.36}$$

The limits of variation of the index m ($0 \leq m \leq 2n$) do not change.

§8. *Transformations of the Asymptotic Series*

In §6 we gave a procedure by which the coefficients A_{nm} are determined in the sequence: $A_{00}, A_{12}, A_{11}, A_{10}, A_{24}, A_{23}, \ldots$; in other words, for each new value of n we begin with the maximum value of m and determine the coefficients in order of decreasing n. But this is not the only possibility. The recursion relation (6.6) also enables us to determine the coefficients in other sequences. In particular, for a given p we may determine the coefficients of the form $A_{n,2n-p}$ in the order of increasing n, where p = 0,1,2,... [28]. We may even sum the terms of this type over n. In this case the sum over n converges.

If m = 2n, then with allowance for (6.36) the recursion relation (6.6) takes the form

$$2nA_{n,2n} = iUA_{n-1,2(n-1)} \quad . \tag{8.1}$$

This implies that

$$A_{n,2n} = \frac{1}{n!}\left(\frac{iU}{2}\right)^n A_{00} \quad , \tag{8.2}$$

$$\sum_{n=0}^{\infty} A_{n,2n}\rho^{-n}(\ln\rho)^{2n} = A_{00}\exp\frac{iU(\ln\rho)^2}{2\rho} \quad . \tag{8.3}$$

For m = 2n-1 we obtain from (6.6)

$$2nA_{n,2n-1} = 4nA_{n,2n} + iUA_{n-1,2n-3} + D_1A_{n-1,2n-2} \quad . \tag{8.4}$$

To solve this equation we use (8.2) and the substitution

$$A_{n,2n-1} = \frac{B_1}{(n-1)!}\left(\frac{iU}{2}\right)^{n-1} + \frac{B_2}{(n-2)!}\left(\frac{iU}{2}\right)^{n-2} \quad . \tag{8.5}$$

Equating the terms on both sides of (8.4) containing (n-1)! and (n-2)!, we find

$$B_1 = A_{11} \quad , \quad B_2 = \frac{i}{4}A_{00}\sum_j \frac{1}{\beta_j^2}\cdot\frac{\partial W}{\partial\theta_j}\cdot\frac{\partial U}{\partial\theta_j} \quad , \tag{8.6}$$

where A_{11} is defined by (6.29) and (7.14) and (7.32) are used to define Δ^* and D_1.

Equation (8.5) then implies

$$\sum_{n=1}^{\infty} A_{n,2n-1}\rho^{-n}(\ln\rho)^{2n-1} = \left(B_1\frac{\ln\rho}{\rho} + B_2\frac{(\ln\rho)^3}{\rho^2}\right)\exp\frac{iU(\ln\rho)^2}{2\rho} \quad . \tag{8.7}$$

Note that the argument of the exponential in (8.7) is the same as in (8.3).

In the general case when m = 2n-p, it is convenient to make the substitution, analogous to (8.5),

$$A_{n,2n-p} = \sum_t B_{t,2t-p} \left(\frac{iU}{2}\right)^{n-t} \frac{1}{(n-t)!} \quad . \tag{8.8}$$

Substituting (8.8) into (6.6) and equating terms with the same values of n, we obtain a recursion relation for the coefficients B_{ts}. It is more unwieldy than (6.6) and so is not given here. The analysis of this recursion relation shows that B_{ts} is nonzero only when

$$t \geq 0 \quad , \quad 0 \leq s \leq \left[\frac{3}{2}\, t\right] \quad , \tag{8.9}$$

where [ ] denotes the integral part.

From (8.8) it follows that

$$\sum_{nm} A_{nm} \rho^{-n} (\ell n\ \rho)^m = \exp\left(\frac{iU(\ell n\ \rho)^2}{2\rho}\right) \sum_{ts} B_{ts} \rho^{-t} (\ell n\ \rho)^s \quad . \tag{8.10}$$

Thus, summing over n for a given p is equivalent to expressing the double series in a different form. We also obtain a recursion relation for the coefficients B_{ts} by substituting into the Schrödinger equation a series in which the left side of Eq. (8.10) is replaced by its right side. It is seen by comparing (6.36) and (8.9) that in the sum containing B_{ts} the number of logarithmic terms for a given power of ρ is smaller. However, the recursion relation is more complicated. Besides $B_{t-1,s}$ there appear also the terms $B_{t-2,s}$ and $B_{t-3,s}$.

We note that the argument of the exponential in (8.10) is purely imaginary. It is possible to obtain an analytical expression for $B_{2t,3t}$ and sum the corresponding terms over t, which again leads to the separation of a phase factor. Thus, in the cases in question summing the terms equally far from the upper limit of the second index is equivalent to separating out a phase factor.

A complete separation of the amplitude and phase of the wave function is obtained by representing the wave function in the form

$$\Psi = R \exp (iS/\hbar) \quad , \qquad (8.11)$$

where R and S are real.

By substituting (8.11) into (7.27) we obtain the equations

$$\left(\frac{\partial S}{\partial r}\right)^2 + \frac{1}{r^2} \sum_{i=1}^{M} \frac{1}{\beta_i^2} \left(\frac{\partial S}{\partial \theta_i}\right)^2 = 2E + \frac{2Z(\hat{\Omega})}{r} + \hbar^2 \frac{\Delta R}{R} \quad , \qquad (8.12)$$

$$\frac{1}{r^M} \cdot \frac{\partial}{\partial r} \left(r^M R^2 \frac{\partial S}{\partial r}\right) + \frac{1}{r^2} \sum_{i=1}^{M} D_i^{(M)} \left(R^2 \frac{\partial S}{\partial \theta_i}\right) = 0 \quad , \qquad (8.13)$$

where $M = 3N-1$ and $D_i^{(M)}$ is the differential operator

$$D_i^{(M)} y = \frac{1}{\beta_i^2 (\sin \theta_i)^{M-i}} \cdot \frac{\partial}{\partial \theta_i} [(\sin \theta_i)^{M-i} y] \quad . \qquad (8.14)$$

Equation (8.13) is the same as the flux conservation equation in classical mechanics, and when $\hbar \to 0$, Eq. (8.12) goes over to the Hamilton-Jacobi equation. The transition to the semiclassical case is considered in more detail in Chapter VII.

Substituting into (8.12) and (8.13) the expansions

$$R(r,\hat{\Omega}) = r^{-M/2} \sum_{nm} R_{nm}(\hat{\Omega}) \rho^{-n} (\ln \rho)^m \quad , \qquad (8.15)$$

$$S(r,\hat{\Omega}) = \rho + W(\hat{\Omega}) \ln \rho + \sum_{nm} S_{nm}(\hat{\Omega}) \rho^{-n} (\ln \rho)^m \quad , \qquad (8.16)$$

where

$$\rho = \kappa r \quad , \quad \kappa = \sqrt{2E} \quad , \quad W = Z/\sqrt{2E} \quad , \qquad (8.17)$$

we get recursion relations for the coefficients R_{nm} and S_{nm}.

These relations are nonlinear and more complicated than (6.6), but once we have R_{00} and S_{00} we can also determine all subsequent terms of the series. We quote the expressions for the first coefficients:

$$R_{12} = 0 \ , \ R_{11} = \frac{1}{2R_{00}} \sum_{i=1}^{M} D_i^{(M)} \left(R_{00}^2 \frac{\partial W}{\partial \theta_i} \right) \ , \qquad (8.18)$$

$$R_{10} = R_{11} - \frac{1}{2} WR_{00} + \frac{1}{2R_{00}} \sum_{i=1}^{M} D_i^{(M)} \left(R_{00}^2 \frac{\partial S_{00}}{\partial \theta_i} \right) \ , \qquad (8.19)$$

$$S_{12} = \frac{1}{2} U \ , \ S_{11} = U + \sum_{i=1}^{M} \frac{1}{\beta_i^2} \cdot \frac{\partial W}{\partial \theta_i} \cdot \frac{\partial S_{00}}{\partial \theta_i} \ , \qquad (8.20)$$

$$S_{10} = \frac{1}{2} W^2 + S_{11} + \frac{1}{2} \sum_{i=1}^{M} \frac{1}{\beta_i^2} \left(\frac{\partial S_{00}}{\partial \theta_i} \right)^2 - \frac{\hbar^2}{R_{00}} \left(\Delta^* - \frac{M(M-2)}{4} \right) R_{00} \ . \qquad (8.21)$$

Note that Eqs. (8.17) used here differ from Eqs. (7.28). The changes were introduced so as to show more explicitly the dependence of the coefficients on $\hbar$ and hence to show more clearly the transition to the semiclassical approximation. This we obtain by setting $\hbar = 0$ in Eqs. (8.12)-(8.21).

The application of the form (8.11) to the radial equation for potential scattering is called the phase-amplitude method [29]. This method is to be distinguished from the method of phase functions [30], in which the form of the wave function for L = 0 is similar to (8.11). In the method of phase functions only the real regular solution to the radial equation is considered, which for L = 0 is given in the form

$$F = a(r) \sin[\phi(r)/\hbar] \ . \qquad (8.22)$$

A supplementary condition is required here, which is chosen to be

$$\frac{dF}{dr} = \frac{\kappa}{\hbar} a(r)\cos[\phi(r)/\hbar] \quad . \tag{8.23}$$

For appropriately chosen boundary conditions, R and S coincide with a and ϕ for $r \to \infty$, but they differ from one another at finite distances. The advantage of the amplitude R (i.e., the amplitude of the wave function) and phase S is that at large distances they vary monotonically and are well approximated by an asymptotic expansion. On the other hand, the amplitude a oscillates about the value $a(\infty)$ with a decreasing oscillation amplitude as $r \to \infty$, and the phase ϕ changes almost discontinuously, oscillating about the phase S. The oscillations in a and ϕ occur twice as fast as in the wave function F.

In the review article of Hull and Breit [29] the first six to eight coefficients of the asymptotic expansion of the amplitude R and phase S in inverse powers of r are given for the Coulomb radial functions. In this case there are no powers of $\ell n\, r$. It turns out that the difference between the quantum and semiclassical results for the radial wave function appears in the amplitude in the r^{-3} term and in the phase in the r^{-2} term. From (8.21) we see that the factor $\hbar^2$ already appears in the coefficient of the r^{-1} term in the phase expansion, where it is found in the centrifugal term, which is retained in the semiclassical approximation for the radial function.

§9. *Various Forms for the Logarithmic Part of the Phase*

From the results of the previous sections the first term of the asymptotic expansion of the wave function for a system of N+1 particles can be written in the form

$$\Psi \sim r^{-(3N-1)/2} A_{00}(\hat{\Omega}) \exp\left\{\frac{i}{\hbar}\left[\sqrt{2E}\, r + \gamma(r,\hat{\Omega})\right]\right\} \quad , \tag{9.1}$$

where the logarithmic part of the phase is

$$\gamma(r,\hat{\Omega}) = \frac{Z(\hat{\Omega})}{\sqrt{2E}} \ln \kappa r \quad . \tag{9.2}$$

Here E denotes the energy in the center-of-mass system, and r is the configuration-space radius vector in the same system, defined by Eq. (7.29). We shall assume the quantity κ in (9.2) to be defined by (7.28).

The form (9.2) for the logarithmic part of the phase is not the only one possible. In fact, the phase of the scattering amplitude A_{00} in the general case was not defined earlier. In order that A_{00} have the meaning of an amplitude, the absolute value of the exponential in (9.1) should be equal to unity. Hence, if Ψ is known, then $|A_{00}(\hat{\Omega})|$ is unambiguously defined. But Ψ, clearly, does not change if we simultaneously replace A_{00} and γ by $\tilde{A}_{00}$ and $\tilde{\gamma}$, where

$$\tilde{A}_{00} = A_{00} \exp\left[-\frac{i}{\hbar}\, y(\hat{\Omega})\right] \quad , \tag{9.3}$$

$$\tilde{\gamma} = \gamma + y(\hat{\Omega}) \quad . \tag{9.4}$$

Here $y(\hat{\Omega})$ is some real function. Thus, the logarithmic phase distortion in the general case is defined up to an arbitrary real function of $\hat{\Omega}$.

We note that the recursion relation (6.6) is changed under the transformation (9.3) and (9.4). We get the recursion relation for the coefficients $\tilde{A}_{nm}$ by substituting into (6.6)

$$A_{nm} = \tilde{A}_{nm} \exp\left[\frac{i}{\hbar}\, y(\hat{\Omega})\right] \quad . \tag{9.5}$$

Let us consider several choices of the function y. If we set

$$y = \frac{Z(\hat{\Omega})}{\sqrt{2E}} \ln 2 \quad , \tag{9.6}$$

we obtain

$$\tilde{\gamma} = \frac{Z(\hat{\Omega})}{\sqrt{2E}} \ln 2\kappa r \quad . \tag{9.7}$$

Here the argument of the logarithm agrees with that usually used in the asymptotic form of the Coulomb wave function for potential scattering. Similarly, by choosing the appropriate value for $y(\hat{\Omega})$, we may replace $\ln \kappa r$ by $\ln r$, $\ln \kappa^2 r$, $\ln Er$, etc. The quantity κ may also be defined by (7.28) or (8.17). Here it should be remembered that the recursion relation (6.6) is valid for $\hbar \neq 1$, if ρ and κ are defined by (7.28). The use of (8.17) leads to a regrouping of the logarithmic terms and coefficients of the expansion which is not exhausted by the transformation (9.5).

A more significant transformation of (9.2) is obtained by replacing $\ln \kappa r$ by such terms as $\ln r_1$, $\ln r_2$, and $\ln r_{12}$. Indeed, the ratios r_1/r, r_2/r, and r_{12}/r are functions only of $\hat{\Omega}$. For example, in the case considered in §6 of the motion of two electrons in the field of an infinitely heavy nucleus (here $\hat{\Omega}$ denotes the coordinates α and θ), we have

$$r_1/r = \cos\alpha \, , \; r_2/r = \sin\alpha \, , \; r_{12}/r = \sqrt{1 - \cos\theta \sin 2\alpha} \; . \tag{9.8}$$

Furthermore we note that the expression (6.2) for $Z(\hat{\Omega})$ can be written as

$$Z(\hat{\Omega}) = \sqrt{2E}\left(\frac{1}{v_1} + \frac{1}{v_2} - \frac{1}{v_{12}}\right) \quad , \tag{9.9}$$

where

$$v_1 = \sqrt{2E}\,\frac{r_1}{r} \, , \; v_2 = \sqrt{2E}\,\frac{r_2}{r} \, , \; v_{12} = \sqrt{2E}\,\frac{r_{12}}{r} \quad . \tag{9.10}$$

The expression (9.2) takes the form

$$\gamma = \left(\frac{1}{v_1} + \frac{1}{v_2} - \frac{1}{v_{12}}\right) \ell n\ \kappa r \quad . \tag{9.11}$$

If we now use

$$y = \frac{1}{v_1}\ \ell n\ \frac{\hbar v_1}{2E} + \frac{1}{v_2}\ \ell n\ \frac{\hbar v_2}{2E} - \frac{1}{v_{12}}\ \ell n\ \frac{\hbar v_{12}}{2E} \quad , \tag{9.12}$$

which is permissible because v_1, v_2, and v_3 depend only on α and θ, then we find

$$\tilde{\gamma} = \frac{\ell n\ r_1}{v_1} + \frac{\ell n\ r_2}{v_2} - \frac{\ell n\ r_{12}}{v_{12}} \quad . \tag{9.13}$$

A form similar to (9.13) is valid also in the general case of a system of charged particles. To demonstrate this, it is convenient to introduce the variables $\underset{\sim}{v}_i$ and t defined by the equations

$$\underset{\sim}{r}_i = \underset{\sim}{v}_i t \quad , \tag{9.14}$$

$$2Et^2 = r^2 \equiv \sum_{i=1}^{N} q_i^{\ 2} \quad , \tag{9.15}$$

where the $\underset{\sim}{q}_i$ are given in (7.19). In (9.14) and (9.15) the center-of-mass system is assumed, where R = 0. It follows from (9.14) and (9.15) with (7.26) that $\underset{\sim}{v}_i$ can be written in the form

$$\underset{\sim}{v}_i = \frac{1}{r} \sum_j c_{ij} \underset{\sim}{q}_j \quad , \tag{9.16}$$

where the c_{ij} are constants. By expressing $\underset{\sim}{q}_j$ in hyperspherical coordinates (7.1) and (7.2), it is easy to see that $\underset{\sim}{v}_i$ depends only on the direction $\hat{\Omega}$ in the 3N-dimensional space of the vectors $\underset{\sim}{q}_1, \ldots, \underset{\sim}{q}_N$.

For the free motion of classical particles emerging from the center of mass at the time t = 0, the quantities $\underset{\sim}{v}_i$ and t have the meaning of the classical velocities and time. If the potential is V = 0, then E = T and (9.15) follows from (9.14) when (7.23) and (7.24) are taken into account. In the scattering of interacting particles, $\underset{\sim}{v}_i$ and t, defined by (9.14) and (9.15), agree with the classical velocities and time in the limit of large distances, for which the main portions of the trajectories lie in the region of almost-free motion. In this sense the dimensions of the interaction region of the particles and the duration of the interaction can be considered small.

Equations (7.18), (9.14), and (9.15) imply

$$Z(\hat{\Omega}) \equiv -rV = -\sqrt{2E}\sum_{j<i}\frac{\zeta_i\zeta_j}{|\underset{\sim}{v}_i-\underset{\sim}{v}_j|} \quad . \tag{9.17}$$

Choosing the function y of the form

$$y = -\sum_{j<i}\frac{\zeta_i\zeta_j}{|\underset{\sim}{v}_i-\underset{\sim}{v}_j|}\,\ln\frac{\hbar|\underset{\sim}{v}_i-\underset{\sim}{v}_j|}{2E} \quad , \tag{9.18}$$

we obtain [31]

$$\tilde{\gamma} = -\sum_{j<i}\frac{\zeta_i\zeta_j}{|\underset{\sim}{v}_i-\underset{\sim}{v}_j|}\,\ln|\underset{\sim}{r}_i-\underset{\sim}{r}_j| \quad . \tag{9.19}$$

The most natural form of the logarithmic part of the phase is the form (9.19), especially when the system contains not only charged but also neutral particles. For a neutral particle, $\zeta_i = 0$ and the logarithmic phase shift in the form (9.19) is independent of the coordinates of the neutral particles.

The expression (9.19) corresponds to the result obtained by Dollard [32] in time-dependent scattering theory. Dollard considered a system of N particles in the field of an infinitely heavy nucleus, described by the Hamiltonian

$$H = H_0 + V \quad , \tag{9.20}$$

where

$$H_0 = -\frac{1}{2}\sum_{j=1}^{N}\frac{\Delta_j}{m_j} \quad , \tag{9.21}$$

$$V = \sum_{j=1}^{N}\left\{\frac{\zeta_0\zeta_j}{r_j} + V_{0j}(r_j) + \sum_{s<j}\left[\frac{\zeta_s\zeta_j}{|\underset{\sim}{r}_s-\underset{\sim}{r}_j|} + V_{sj}(\underset{\sim}{r}_s-\underset{\sim}{r}_j)\right]\right\} \quad . \tag{9.22}$$

Here V_{sj} is a square-integrable potential, and the system of units is used in which $\hbar = 1$.

If there is no Coulomb interaction ($\zeta_j = 0$, $j = 1,\ldots,N$), then the operator

$$Q(t) = \exp(iHt)\exp(-iH_0t) \quad , \tag{9.23}$$

acting in the space of square-integrable functions of the 3N variables $\underset{\sim}{r}_1,\ldots,\underset{\sim}{r}_N$, converges strongly to the limiting values $Q^{\pm}$ for $t \to \pm\infty$. Strong convergence means convergence in norm, which is expressed in terms of the integral of the square of the modulus over the coordinates:

$$\lim_{t\to\pm\infty} ||Q(t)f - Q^{\pm}f|| = 0 \quad . \tag{9.24}$$

The limiting values $Q^{\pm}$, which are usually called the wave operators, define the scattering operator

$$S = Q^{+*}Q^{-} \quad . \tag{9.25}$$

In the presence of the Coulomb interaction the operators (9.23) do not converge strongly. This is analogous to the fact that the solution of the stationary problem does not have the asymptotic form of free motion. Dollard showed that the modified operator

$$Q_C(t) = \exp(iHt)\exp[-iH_0t - i\tilde{H}_{0C}(t)] \tag{9.26}$$

converges strongly, where

$$\tilde{H}_{0C}(t) = \text{sign}(t) \sum_{j=1}^{N} \left[\frac{m_j\zeta_0\zeta_j}{(-\Delta_j)^{1/2}} \ln\left(\frac{-2|t|\Delta_j}{m_j}\right) + \right.$$

$$\left. + \sum_{s=1}^{j-1} \frac{m_s m_j \zeta_s \zeta_j}{(-D_{sj})^{1/2}} \ln\left(\frac{-2|t|D_{sj}}{m_s m_j (m_s + m_j)}\right)\right] . \tag{9.27}$$

Here

$$D_{sj} = (m_j\nabla_s - m_s\nabla_j)^2 . \tag{9.28}$$

Since the operator $-i\nabla$ corresponds to momentum and in the asymptotic domain $\underset{\sim}{v}_j$ is equal to velocity, we have the correspondence

$$\nabla_j \to im_j\underset{\sim}{v}_j \quad , \quad \Delta_j \to -m_j^2 v_j^2 . \tag{9.29}$$

Substituting (9.29) into (9.27) and using (9.14), we obtain an expression similar to (9.19). These expressions differ in that the result obtained from (9.27) contains an additional factor under the logarithm sign of twice the relative momentum, but as we showed above, this is not significant because the particle velocities depend only on the direction $\hat{\Omega}$. In addition, the expression (9.27) differs from (9.19) in sign for $t > 0$, because the logarithmic terms in the operator (9.26) and in the asymptotic configuration-space wave function have somewhat different roles. In (9.26) the logarithmic part in the asymptotic domain directly compensates for the Coulomb interaction in H, and therefore the sign of the operator $\tilde{H}_{0C}$ for $t > 0$ should agree with the sign of the Coulomb potential. The logarithmic part of the phase of the configuration-space wave function compensates for the Coulomb interaction when the wave function is substituted into the Schrödinger equation. When the second derivative with respect to r is calculated, a corresponding term appears which, however, has a sign opposite to the sign of the logarithmic part of the phase.

Buslaev and Matveev [33] give a generalization of the operator (9.26) for the case in which the potential falls off more slowly than the Coulomb potential. The phase shift of the configuration-space wave function for the potential $V(r) = -Zr^{-\alpha}$, where $1/2 < \alpha \leq 1$, according to Buslaev and Matveev, can be written in the form

$$\gamma = -\frac{1}{\sqrt{2E}}\int^{r} V(r')dr' \quad . \tag{9.30}$$

For $\alpha = 1$ the expression (9.30) is equivalent to (9.2).

Using the phase shift corresponding to (9.26)-(9.28), Veselova [34] obtained expressions for the scattering amplitudes for two and three charged particles in terms of the T-matrix elements. Veselova showed that the T matrix in the neighborhood of the energy shell has a logarithmically oscillating singularity.

Sakhnovich [35] obtained a generalization of Dollard's result that allows for the possibility of bound states of several particles in a system of many charged particles.

It should be mentioned that the asymptotic forms (9.1) and (9.2) turn out to be valid in the limiting cases for which exact solutions are known. If $Z(\hat{\Omega}) \to 0$, then (9.1) and (9.2) go over to the free-particle asymptotic form applicable for a short-range potential. Another well-known case is $Z(\hat{\Omega}) = \text{const}$. The Coulomb potential in a many-dimensional space, independent of the direction $\hat{\Omega}$, does not have an immediate physical interpretation. However, as a mathematical problem this example can be solved exactly. The multidimensional Coulomb wave functions are considered in §16. Comparison with them and also with the well-known asymptotic form of the three-dimensional Coulomb wave function shows that (9.1) and (9.2) are valid also in this case. We note that the phase shift in the general case turns out to be the same as for a constant Z equal in magnitude to $Z(\hat{\Omega})$ for the direction in question. From the physical point of view this is understandable, for in the final stage of scattering the system moves in the one direction $\hat{\Omega}$ for an infinitely long time.

We may also regard as known the asymptotic form for a system containing two charged particles and one or several neutral particles. The problem is then solved by separation of variables. In the domain where the interaction between the neutral particles may be neglected, the general solution is a superposition of wave functions that describe the relative Coulomb motion of the charged particles, as well as the free motion of the remaining particles and of the center of mass. The case of two charged particles and one neutral particle was studied by Hart, Gray, and Guier [36], who showed that for Eq. (1.1) with the potential

$$V = -\frac{1}{ar_2} \equiv -\frac{1}{ar\sin\alpha}, \tag{9.31}$$

the asymptotic solution contains the logarithmic term

$$\frac{\ln(2\kappa r\sin\alpha)}{\kappa a\sin\alpha} \tag{9.32}$$

in the exponent. Since for the potential (9.31) one has $Z(\hat{\Omega}) = 1/a\sin\alpha$, the logarithmic term (9.32) agrees with (9.2) to within a transformation of the type (9.3) and (9.4).

§10. *Validity of the Asymptotic Forms*

The conditions for the validity of the asymptotic forms (6.39) or (9.1) are essentially the same as in the case of short-range forces. Therefore, let us first look at this case in more detail. The problem that corresponds to the ionization of a hydrogen atom by an electron is that of the motion of two neutral particles of unit mass in the field of a fixed center.

The general solution of the Schrödinger equation that satisfies the boundary conditions of outgoing radial motion can be represented as a superposition of plane waves in the free-motion region:

$$\Psi = \int \delta_+\left(E - \frac{1}{2}p_1^2 - \frac{1}{2}p_2^2\right) a(\underset{\sim}{p}_1, \underset{\sim}{p}_2) \exp(i\underset{\sim}{p}_1 \cdot \underset{\sim}{r}_1 + i\underset{\sim}{p}_2 \cdot \underset{\sim}{r}_2) d\underset{\sim}{p}_1 d\underset{\sim}{p}_2 \ , \tag{10.1}$$

where the δ_+ function is defined in (5.10).

In order to get an asymptotic expression for the integral (10.1), we introduce the hyperspherical coordinates (7.1) and (7.2) into the six-dimensional space of the vectors $\underset{\sim}{p}_1$ and $\underset{\sim}{p}_2$. For the exponential we use the asymptotic expression derived in §15. Using the properties of the δ_+ function, we find that for large r_1 and r_2 the stationary points of the integral (10.1) will be the values of $\underset{\sim}{p}_1$ and $\underset{\sim}{p}_2$ satisfying the conservation of energy $p_1^2 + p_2^2 = 2E$ and corresponding to the direction $\hat{\Omega}_{\underset{\sim}{r}}$. This means that the stationary points correspond to the directions $\underset{\sim}{r}_1$ and $\underset{\sim}{r}_2$ and are proportional to r_1 and r_2 in absolute value. Thus, the stationary values are

$$\underset{\sim}{p}_i = \kappa r^{-1} \underset{\sim}{r}_i \equiv \underset{\sim}{k}_i \quad . \tag{10.2}$$

In this way we find

$$\Psi \sim A(\hat{\Omega}) r^{-5/2} e^{i\kappa r} \quad , \qquad (10.3)$$

where

$$A(\hat{\Omega}) = (-2\pi i)^{5/2} \kappa^{3/2} a(\underset{\sim}{k}_1, \underset{\sim}{k}_2) \quad . \qquad (10.4)$$

Equation (10.3) describes spherically diverging free motion. The existence of a stationary point means that at large distances only that wave is retained in the superposition (10.1) for which the velocity and distance are related by the classical equation of free motion.

In order that the asymptotic form (10.3) be valid, it is necessary first of all that we be able to treat the motion in the asymptotic domain as free, i.e., that we be able to ignore the potential energy

$$|V| << E \quad . \qquad (10.5)$$

However, because (10.3) describes free diverging motion from a point center, it is also necessary that a stronger condition be satisfied. Free motion should occur also throughout much of the region where the separation is not so large. This means that the distance to the asymptotic domain should be large compared to the dimensions of the effective interaction region of the particles.

The point of the Coulomb asymptotic forms (6.39) and (9.1) is that they describe almost free (logarithmically distorted) motion, and, hence, (10.5) is a necessary condition for their validity as well. In the Coulomb case (10.5) takes the form

$$\frac{|Z(\hat{\Omega})|}{r} << E \quad . \qquad (10.6)$$

It does not yet follow, however, that

$$\frac{|Z(\hat{\Omega})|\ln r}{\kappa} << \kappa r \quad . \qquad (10.7)$$

Thus, we may not assert that the logarithmic term is necessarily less in absolute value than the first term in the phase shift. However, since the phase of the wave function is defined only up to a constant, what is important in comparing the roles of the various terms in the phase is not their absolute value, but rather how quickly they vary with r. From (10.6) there follows an inequality for the derivatives of (10.7)

$$\frac{|Z(\hat{\Omega})|}{\kappa r} << \kappa \quad , \qquad (10.8)$$

whence it is seen that the logarithmic term in the phase shift leads to little distortion of the free motion.

Equation (10.6) implies that the domain of validity of the Coulomb asymptotic form depends on the direction $\hat{\Omega}$. The distance r at which the asymptotic form is valid increases with $Z(\hat{\Omega})$. In those directions $\hat{\Omega}$ for which the particles are in contact ($r_1 = 0$ or $r_2 = 0$ or $r_{12} = 0$), $Z(\hat{\Omega})$ goes to infinity along with the potential. In such directions the asymptotic forms are invalid. As $\hat{\Omega}$ approaches these directions, the domain of validity of the asymptotic form is removed to infinity.

More precisely, the domain of validity of the Coulomb asymptotic form is determined by means of the recursion relation (6.6) and the Eqs. (6.28)-(6.30), which follow from (6.6). It must be borne in mind here that the expansion (6.5) involves inverse powers of κr. The dependence of the coefficients $A_{nm}(\hat{\Omega})$ on $Z(\hat{\Omega})$ and κ is determined by the function $W = Z(\hat{\Omega})/\kappa$, which enters into the recursion relation explicitly and also through U, D_1, and D_2. We see from Eqs. (6.7)-(6.9) that the domain of validity of the asymptotic form is determined not only by the first power of $Z(\hat{\Omega})$ [which leads to (10.8)], but also by its square and its derivatives. One must also remember that because of the action of the operators D_1 and D_2 the coefficients of the higher terms in the expansion will contain higher order derivatives of both $Z(\hat{\Omega})$ and $A_{00}(\hat{\Omega})$. Therefore, for the general Coulomb interaction it is difficult to give a concrete quantitative estimate for the boundary of the asymptotic domain. Some idea of this domain is given by the results of numerical calculation for the simplified problem discussed in §14.

From the above discussion it is clear that the asymptotic form cannot be applied, for example, when r_{12} is small. The critical remarks of Temkin [37] were evoked by the neglect of this circumstance. He called attention to the fact that $Z(\hat{\Omega}) = -\infty$ for $r_{12} = 0$ and hence concluded that the asymptotic form (6.39) is not valid because it contains a singularity. In fact, in those regions where the asymptotic form is applicable, it is by its nature nonsingular. If $r \neq 0$, then $r_{12} \to 0$ implies that $\alpha \to \pi/4$ and $\theta_{12} \to 0$. We see from (6.2) that in this case $|Z(\hat{\Omega})|$ indeed increases, but at the same time the distance r at which the asymptotic form is applicable also increases. The radius r increases in such a way that r_{12} remains large, so that in the domain of validity (10.5) and (10.8) are always satisfied. Thus, even for large $|Z(\hat{\Omega})|$ the logarithmic term gives a relatively small correction to the free-particle wave function. At the point $r_{12} = 0$, where $Z(\hat{\Omega}) = -\infty$, the domain of validity of the asymptotic form vanishes (it has been displaced to infinity). In order to apply the asymptotic form (6.39), it is essential that all distances between charged particles be large. Peterkop [38] analyzes Temkin's criticisms [37] in detail.

From the condition (10.5) we see that the asymptotic form (6.39) is inapplicable if $E = 0$. We recall also that the expansion (6.5) involves powers of κr, and that the function W, which plays an essential role in the recursion relation, contains κ in the denominator. Thus, as $E \to 0$ the domain of validity of the asymptotic form is removed to infinity. Physically this means that at zero energy the motion can nowhere be considered as free. An asymptotic form for the case $E = 0$ was proposed by Rudge and Seaton [39], who used the semiclassical approximation (see §27).

§11. *Asymptotic Behavior for Finite r_{12} and Nonspherical Asymptotic Forms*

Asymptotic behavior for finite r_{12}. It was noted in §10 that the asymptotic form (6.39) is applicable only when all distances between particles are large. Thus, it is not valid if r_{12} remains finite. The asymptotic behavior for finite r_1 or r_2 is determined by Eqs. (1.3) and (1.4). We can expect that the asymptotic domain, in which r_1 and $r_2 \to \infty$ and r_{12} remains finite, is of less interest for the ionization problem, because the wave function in this domain should fall off faster than $r^{-5/2}$ owing to the repulsion of the electrons.

The asymptotic behavior for finite r_{12} was described by Peterkop [40]. Let us consider first the asymptotic behavior for a short-range potential. In this case for sufficiently large r_1 and r_2 we may ignore the potentials depending on these coordinates, and the Schrödinger equation takes the form

$$\left(\frac{1}{2}\Delta_1 + \frac{1}{2}\Delta_2 - V(r_{12}) + E\right)\Psi = 0 \quad . \qquad (11.1)$$

This equation separates in terms of the variables

$$\underset{\sim}{R} = (\underset{\sim}{r}_1 + \underset{\sim}{r}_2)/2 \quad , \quad \underset{\sim}{r}_{12} = \underset{\sim}{r}_1 - \underset{\sim}{r}_2 \quad . \tag{11.2}$$

The appropriate solutions are of the form

$$\psi = \exp(i\underset{\sim}{P}\cdot\underset{\sim}{R})\phi(\underset{\sim}{p},\underset{\sim}{r}_{12}) \quad , \tag{11.3}$$

where ϕ is determined from the equation

$$(\Delta_{r_{12}} - V(r_{12}) + p^2)\phi = 0 \quad , \tag{11.4}$$

and the vectors $\underset{\sim}{P}$ and $\underset{\sim}{p}$ are related by

$$\frac{1}{4} P^2 + p^2 = E \quad . \tag{11.5}$$

The exponential in (11.3) describes the free motion of the center of mass of the particles, and ϕ describes their relative motion. We assume that ϕ is chosen to behave asymptotically as an "incident wave + converging wave" --

$$\phi \sim \exp(i\underset{\sim}{p}\cdot\underset{\sim}{r}_{12}) + \frac{a^*(-\hat{\Omega}_{12})}{r_{12}} e^{-ipr_{12}} \quad , \tag{11.6}$$

where $a(\hat{\Omega}_{12})$ is the scattering amplitude for the potential $V(r_{12})$.

The general solution to Eq. (11.1) with the diverging-wave boundary condition can be represented as a superposition

$$\Psi = \int \delta_+\left(E - \frac{1}{4} P^2 - p^2\right) b(\underset{\sim}{P},\underset{\sim}{p}) \exp(i\underset{\sim}{P}\cdot\underset{\sim}{R})\phi(\underset{\sim}{p},\underset{\sim}{r}_{12}) d\underset{\sim}{P} d\underset{\sim}{p} \quad . \tag{11.7}$$

By substituting (11.6) into (11.7) and making use of the properties of the δ_+ function, we can demonstrate that at large r_{12} the superposition (11.7) is equal to (10.1). Here, we have

$$b(\underset{\sim}{P},\underset{\sim}{p}) = a(\underset{\sim}{p}_1,\underset{\sim}{p}_2) \quad , \tag{11.8}$$

where

$$\underset{\sim}{P} = \underset{\sim}{p}_1 + \underset{\sim}{p}_2 \quad , \quad \underset{\sim}{p} = (\underset{\sim}{p}_1 - \underset{\sim}{p}_2)/2 \quad . \tag{11.9}$$

Equations (11.2) and (11.9) imply

$$\frac{1}{4}\, P^2 + p^2 = \frac{1}{2}\, p_1^2 + \frac{1}{2}\, p_2^2 \, , \; \underset{\sim}{P}\cdot\underset{\sim}{R} + \underset{\sim}{p}\cdot\underset{\sim}{r}_{12} = \underset{\sim}{p}_1\cdot\underset{\sim}{r}_1 + \underset{\sim}{p}_2\cdot\underset{\sim}{r}_2 \, , \; d\underset{\sim}{P}d\underset{\sim}{p} = d\underset{\sim}{p}_1 d\underset{\sim}{p}_2 . \tag{11.10}$$

Equation (11.7) is more general than (10.1), because it is valid also for small r_{12}.

If r_{12} is finite and $R \to \infty$, then for the exponential in (11.7) we may use the three-dimensional version of Eq. (15.10). The stationary point in the integration over $\underset{\sim}{P}$ occurs for

$$\underset{\sim}{P} = 2\sqrt{E - p^2}\; R^{-1}\underset{\sim}{R} \equiv \underset{\sim}{K} \quad . \tag{11.11}$$

The asymptotic expression (11.7) takes the form

$$\Psi \sim -\frac{4\pi i}{R} \int b(\underset{\sim}{K},\underset{\sim}{p})\,\phi(\underset{\sim}{p},\underset{\sim}{r}_{12})\,e^{iKR}d\underset{\sim}{p} \quad . \tag{11.12}$$

To this integral we can also apply the method of stationary phase. The stationary point is determined by the condition $dK/dp = 0$, whence follows $p_{st} = 0$. We integrate over p in the range $0 \le p \le \sqrt{E}$, because for $p > \sqrt{E}$ the quantity $K = 2\sqrt{E - p^2}$ becomes imaginary and $\exp(iKR)$ quickly falls off. From the equations for the asymptotic behavior of Fourier integrals [10] we find

$$\int_0^{\sqrt{E}} F(p)\,e^{iKR}\,p^2 dp \sim \frac{E^{3/4}\pi^{1/2}}{4i^{3/2}R^{3/2}}\, F(0)\,e^{2i\sqrt{ER}} \quad . \tag{11.13}$$

To derive Eq. (11.13) we first transform to an integration over K. Then the phase of the exponential is no longer stationary, and the integrand increases in the neighborhood of $p = 0$. To remove the zero at $p = 0$, we integrate by parts. Of the integrals obtained, the main one is of the type

$$I = \int_\alpha^\beta g(t)e^{ixt}(t-\alpha)^{\lambda-1}(\beta-t)^{\mu-1}dt \ , \ 0 < \lambda \ , \ \mu \le 1 \quad . \tag{11.14}$$

The asymptotic expression for this integral can be written as the difference of two functions that involve the limits of integration:

$$I = B(x) - A(x) \quad . \tag{11.15}$$

For $x \to \infty$ the asymptotic expansions of the functions A and B have the following leading terms [10]:

$$A \sim -g(\alpha)(\beta-\alpha)^{\mu-1}\Gamma(\lambda)(-ix)^{-\lambda}e^{ix\alpha} \quad , \tag{11.16}$$

$$B \sim g(\beta)(\beta-\alpha)^{\lambda-1}\Gamma(\mu)(ix)^{-\mu}e^{ix\beta} \quad . \tag{11.17}$$

In raising $-i$ and i to any power it is necessary to take the argument with lowest absolute value [see (2.19)].

By substituting (11.13) into (11.12), recalling that $\phi(\underset{\sim}{p},\underset{\sim}{r})$ is spherically symmetric for $p = 0$ (contains only the s component), and expressing R and E in terms of r and κ, we obtain

$$\Psi \sim (-2\pi i)^{5/2}\kappa^{3/2}b(\underset{\sim}{K},0)\phi(0,r_{12})r^{-5/2}e^{i\kappa r} \quad . \tag{11.18}$$

If we replace $b(\underset{\sim}{K},0)$ in (11.18) by the more general form $a(\underset{\sim}{k}_1,\underset{\sim}{k}_2)$, where $\underset{\sim}{k}_i$ is defined by (10.2), then by virtue of (11.6) Eq. (11.18) for $r_{12} \to \infty$ goes over to (10.3) and (10.4). Thus, (11.18) represents a generalization of the asymptotic form (10.3) and (10.4), valid for both large and small r_{12}.

Let us now consider the Coulomb potential. Using the well-known expansion

$$\frac{1}{|\underset{\sim}{r}_1-\underset{\sim}{r}_2|} = \frac{1}{r_1}\sum_{\ell=0}^{\infty} P_\ell(\cos\theta_{12})\left(\frac{r_2}{r_1}\right)^\ell \ , \ r_1 > r_2 \ , \tag{11.19}$$

we obtain for $R > r_{12}/2$

$$\frac{1}{r_1} + \frac{1}{r_2} = \frac{1}{|\underset{\sim}{R} + \frac{1}{2}\underset{\sim}{r}_{12}|} + \frac{1}{|\underset{\sim}{R} - \frac{1}{2}\underset{\sim}{r}_{12}|} = \frac{2}{R}\sum_{n=0}^{\infty} P_{2n}(\cos\theta)\left(\frac{r_{12}}{2R}\right)^{2n} , \qquad (11.20)$$

where θ is the angle between $\underset{\sim}{R}$ and $\underset{\sim}{r}_{12}$.

Taking into account only the first term of the series (11.20) (the next term falls off as $r_{12}^2 R^{-3}$), we obtain instead of (11.7) an expression in which $\exp(i\underset{\sim}{P}\cdot\underset{\sim}{R})$ is replaced by the Coulomb wave function that satisfies the equation

$$\left(\Delta_R + \frac{8}{R} + P^2\right)\psi(\underset{\sim}{P},\underset{\sim}{R}) = 0 \quad . \qquad (11.21)$$

We assume that in analogy to (11.6) ψ behaves asymptotically like an incident wave + converging wave. Using the asymptotic expression for the Coulomb wave function (see §16), we obtain an expression similar to (11.12):

$$\Psi \sim -\frac{4\pi i}{R}\int b(\underset{\sim}{K},\underset{\sim}{p})\phi(\underset{\sim}{p},\underset{\sim}{r}_{12})e^{iKR+(4i/K)\ln 2KR}d\underset{\sim}{p} \quad . \qquad (11.22)$$

The main difference between this and (11.12) is the logarithmic term in the exponential. In addition, ϕ here denotes the Coulomb wave function for the relative motion of electrons.

Applying the method of stationary phase to (11.22), we obtain the analog of Eq. (11.18)

$$\Psi \sim (-2\pi i)^{5/2}\kappa^{3/2}r^{-5/2}e^{i\kappa r+(i\sqrt{8}/\kappa)\ln 2\kappa r}\int b(\underset{\sim}{K},\hat{\Omega}_p)\phi(\hat{\Omega}_p,\underset{\sim}{r}_{12})\frac{d\hat{\Omega}_p}{4\pi} ,$$
$$p = 0 \quad . \qquad (11.23)$$

Since the Coulomb wave function is not spherically symmetric at zero energy, in contrast with (11.18), we retain the integral over $\hat{\Omega}_p$ in (11.23). We find here also the characteristic logarithmic phase shift. The coefficient $\sqrt{8}$ is due to the interaction of the electrons with the nucleus for

$$\frac{1}{r_1} + \frac{1}{r_2} = \frac{\sqrt{8}}{r} \quad , \text{ if } \quad r_1 = r_2 \quad . \tag{11.24}$$

Electron interaction effects are described by the function ϕ.

For Coulomb repulsion between particles the wave function ϕ, normalized according to (11.6), falls off exponentially at finite distances when $p \to 0$. It is to be expected also that owing to the repulsion of the electrons the amplitude b goes to zero for $p = 0$. Thus, Eq. (11.23) indicates that for finite r_{12} the wave function falls off faster than $R^{-5/2}$ for $R \to \infty$ and, hence, does not contribute to the flux or ionization cross section.

Note that in applying the method of stationary phase to (11.22), we brought the factor $\exp[(4i/K)\ln R]$ through the integral sign with the value of K corresponding to the stationary point, i.e., we treated $\ln R$ as a finite quantity. This is justified by the slow growth of $\ln R$ in comparison with R. The correctness of this procedure has been demonstrated more rigorously using the method of stationary phase by Riekstyn'sh [11].

The somewhat different case of the asymptotic behavior of a three-particle wave function with a finite distance between one pair of particles was discussed by Doolen and Nuttall [41]. They examined the asymptotic behavior of the wave function describing the ionization of a hydrogen atom by a neutral particle with the mass of an electron, in the domain where the neutral particle is far from the nucleus and the electron is close to it. In this case an attractive force is felt, and bound states exist. The main contribution to the asymptotic behavior ($\sim$ const + r^{-1}) is given by the sum in

(1.3) over discrete states. Doolen and Nuttall [41] estimated the contribution from the integral over the continuous spectrum in (1.3) to be $\sim r^{-2}$. Their result corresponds to the Born approximation for the collision of an electron with a hydrogen atom, for in the Born approximation the incident (scattered) electron is described by a plane wave. The derivation of this result consists essentially of obtaining from an expression of the form (11.7) an expression similar to (11.18) but containing r^{-2} instead of $r^{-5/2}$. This difference arises because in the derivation of (11.18) the potential $V(r_{12})$ is assumed to be short range, whereas Doolen and Nuttall considered the Coulomb potential. For a Coulomb attraction the wave function $\phi(\underset{\sim}{p},\underset{\sim}{r}_{12})$ goes to infinity as $p^{-1/2}$ for $p \to 0$. It can be shown that in this case the amplitude $b(\underset{\sim}{K},\underset{\sim}{p})$ also has such a singularity. Consequently, a factor $\sim p^{-1}$ appears in the integrand of (11.12), and since $p = 0$ is a stationary point, the method of stationary phase in this case leads to an expression that falls off more slowly.

Nonspherical asymptotic forms. The asymptotic form considered in this chapter represents a spherical wave in multidimensional configuration space. Nonspherical asymptotic forms are also possible. The general asymptotic form for a system of two electrons in the field of a fixed hydrogen nucleus can be written as

$$\psi = r^{-n_o} e^{iS^{(o)}+i\gamma} \sum_{n\geq 0} \sum_{m=0}^{2n} A_{nm}(\hat{\Omega}) \frac{(\ln r)^m}{r^n} , \qquad (11.25)$$

where $S^{(o)}$ is a solution of the Hamilton-Jacobi equation for free particles

$$(\nabla_1 S^{(o)})^2 + (\nabla_2 S^{(o)})^2 = \kappa^2 , \qquad (11.26)$$

which can be reduced to the form $S^{(o)} = rf(\hat{\Omega})$. The logarithmic phase factor is

$$\gamma = \frac{1}{k_1} \ln(k_1 r_1 + \underset{\sim}{k}_1 \underset{\sim}{r}_1) + \frac{1}{k_2} \ln(k_2 r_2 + \underset{\sim}{k}_2 \underset{\sim}{r}_2) - $$

$$- \frac{1}{k_{12}} \ln(k_{12} r_{12} + \underset{\sim}{k}_{12} \underset{\sim}{r}_{12}) \quad . \tag{11.27}$$

The vectors $\underset{\sim}{k}_1$ and $\underset{\sim}{k}_2$ are defined as

$$\underset{\sim}{k}_1 = \nabla_1 S^{(o)} \quad , \quad \underset{\sim}{k}_2 = \nabla_2 S^{(o)} \quad , \quad \underset{\sim}{k}_{12} = \underset{\sim}{k}_1 - \underset{\sim}{k}_2 \quad . \tag{11.28}$$

The phase γ satisfies

$$[(\nabla_1 S^{(o)})\nabla_1 + (\nabla_2 S^{(o)})\nabla_2]\gamma = -V \quad , \tag{11.29}$$

where the potential energy V is determined by (1.2). Another form of γ can be obtained if one changes the signs of $\underset{\sim}{k}_1$ and $\underset{\sim}{k}_2$ and takes the complex conjugate wave function.

The use of Eq. (11.26) allows terms $\sim r^{-n_o}$ to be cancelled. Equation (11.29) makes possible the cancellation of terms $\sim r^{-n_o - 1} \ln r$ and the elimination of the potential energy, if (11.25) is substituted in the Schrödinger equation (1.1). Terms containing $r^{-n_o - 1}$ lead to the equation

$$\sum_{j=1}^{2} \left[2r\nabla_j S^{(o)} \left(\nabla_j - \frac{n_o}{r^2} \underset{\sim}{r}_j \right) + r\Delta_j S^{(o)} \right] A_{00}(\hat{\Omega}) = 0 \ . \tag{11.30}$$

At points where $S^{(o)}$ reaches the maximum value $S^{(o)} = \kappa r$, $\nabla S^{(o)}$ as a six-dimensional vector is parallel to $\underset{\sim}{r}$ and orthogonal to $\nabla A_{00}(\hat{\Omega})$. Then the first term of (11.30) vanishes and the remaining two give the expression

$$n_o = \frac{r}{2\kappa} (\Delta_1 + \Delta_2) S^{(o)} \quad , \tag{11.31}$$

which is valid at points where $S^{(o)} = \kappa r$.

Recurrence relations can be obtained which determine other A_{nm} if A_{00} is known. A_{00} may depend in an arbitrary way on those angles of $\hat{\Omega}$ of which $S^{(o)}$ is independent.

The spherical asymptotic form is obtained by putting $S^{(o)} = \kappa r$. Then $n_o = 5/2$ and γ is essentially equal to (9.13). Examples of nonspherical forms are

$$S^{(o)} = \underset{\sim}{k}_1 \underset{\sim}{r}_1 + \underset{\sim}{k}_2 \underset{\sim}{r}_2 \quad , \quad n_o = 0 \quad , \tag{11.32}$$

$$S^{(o)} = \underset{\sim}{k}_1 \underset{\sim}{r}_1 + k_2 r_2 \quad , \quad n_o = 1 \quad , \tag{11.33}$$

$$S^{(o)} = \frac{1}{2} \underset{\sim}{K}(\underset{\sim}{r}_1 + \underset{\sim}{r}_2) + \frac{1}{2} k_{12} r_{12} \quad , \quad n_o = 1 \quad , \tag{11.34}$$

$$S^{(o)} = k_1 r_1 + k_2 r_2 \quad , \quad n_o = 2 \quad . \tag{11.35}$$

It is assumed that $k_1^{\;2} + k_2^{\;2} = \kappa^2$ and $K^2 + k_{12}^2 = 2\kappa^2$.

Equation (11.32) yields a plane-wave asymptotic form. In this case A_{00} does not depend on Ω and the expansion (11.25) does not contain powers of $\ell n\ r$. The expression for γ in the plane-wave case agrees with the form given by Redmond (quoted by Rosenberg [176]).

The forms (11.33)-(11.35) correspond to one or two binary collisions [58,59].

CHAPTER III

THE METHOD OF K HARMONICS IN THE IONIZATION PROBLEM

§12. *The Method of K Harmonics*

In order to carry out numerical calculations the wave function is usually represented as a series involving some system of functions. In the excitation of discrete levels an expansion with respect to the eigenfunctions of the atom is used. If such a method is used to treat ionization processes also, then the unknown wave function can be written in the form

$$\Psi = \sum_{n\ell m} F_{n\ell m}(\underset{\sim}{r}_1)\phi_{n\ell m}(\underset{\sim}{r}_2) + \sum_{\ell m} \int_0^\infty F_{\varepsilon\ell m}(\underset{\sim}{r}_1)\phi_{\varepsilon\ell m}(\underset{\sim}{r}_2)d\varepsilon \quad , \qquad (12.1)$$

where $\phi_{n\ell m}$ is an eigenfunction from the discrete spectrum of the atom and $\phi_{\varepsilon\ell m}$ is from the continuous spectrum. When exchange is taken into account, Eq. (12.1) has to be symmetrized or antisymmetrized.

An expansion of the form (12.1) is inconvenient in that the parameter ε varies continuously. It leads to a system of integrodifferential equations containing both discrete and continuous sets of unknown functions, which poses difficulties for numerical calculations. Expansions of the form (12.1) taking into account the continuous spectrum have not as yet been used in applied calculations.

In order to obtain a problem containing only a discrete set of unknown functions, it is expedient to represent the part of (12.1) involving the continuous spectrum as a series of so-called K harmonics. The K harmonics, or hyperspherical harmonics, are the eigenfunctions of the Laplace operator on the unit hypersphere in the multidimensional space. According to (7.15) the K harmonics in six-dimensional space satisfy the equation

$$[\Delta^* + K(K+4)]Y_{K\gamma}(\hat{\Omega}) = 0 \quad , \tag{12.2}$$

where

$$K = 0,1,2,\ldots \quad . \tag{12.3}$$

The index γ denotes the set of remaining quantum numbers specifying the eigenfunction.

The K harmonics form a complete set of functions in the space $\hat{\Omega}$. They can be chosen to be real and orthonormal:

$$\int Y_{K\gamma}Y_{K'\gamma'}d\hat{\Omega} = \delta_{KK'}\delta_{\gamma\gamma'} \quad . \tag{12.4}$$

We can replace the expansion (12.1) by

$$\Psi = \sum_{n\ell m} F_{n\ell m}(\underset{\sim}{r}_1)\phi_{n\ell m}(\underset{\sim}{r}_2) + r^{-5/2}\sum_{K\gamma} F_{K\gamma}(r)Y_{K\gamma}(\hat{\Omega}) \quad . \tag{12.5}$$

The factor $r^{-5/2}$ has been introduced to simplify the system of radial equations.

The quantities represented by $\hat{\Omega}$ determine the directions of the two particles and the ratio of their distances. In the asymptotic domain the latter is proportional to the ratio of the velocities. Thus, each K harmonic describes a state with some definite angular distribution of the particles as well as with a specific velocity, or energy, distribution.

We obtain a system of equations for the functions $F_{n\ell m}$ and $F_{K\gamma}$ by, as usual, substituting (12.5) into the Schrödinger equation, multiplying on the left by $\phi^*_{n'\ell'm'}(\underset{\sim}{r}_2)$ or $Y_{K'\gamma'}(\hat{\Omega})$, and integrating over $\underset{\sim}{r}_2$ or $\hat{\Omega}$. The part of the system of equations corresponding to the K harmonics is reduced further to (13.1). It should be noted that the K harmonics are not orthogonal to the atomic wave functions, and consequently the unknown functions in (12.5) are not uniquely determined. This circumstance is relevant, of course, only for finite distances, because the atomic wave functions fall off rapidly at large distances. To remove the ambiguity we can introduce supplementary conditions. In this regard the situation is analogous to the case of exchange, when an expansion is also performed over mutually overlapping sets of functions (see §5).

Following the work of Simonov and Badalyan [42,43], K harmonics were widely applied in calculations of discrete nuclear states. A review of recent work is given by Baz' *et al.* [44]. The use of K harmonics to take into account the continuous spectrum in ionization problems is discussed by several authors [45-47]. An expansion of the form (12.5) was first applied to a collision problem by Delves [48], who calculated the cross section for the disintegration of a deuteron by a neutron. In this case the discrete spectrum has only one state. Delves' calculation is quite approximate, since in the sum over K harmonics only one harmonic (K = 0) is included. Until very recently K harmonics had not been applied in calculations of electron-atom collisions.

Several different sets of K harmonics are known, because Eq. (12.2) admits separation of variables in several different coordinate systems. When all harmonics of a given K are taken into account, the various sets are equivalent, since the harmonics of different sets with the same index K can be expressed as linear combinations of one another.

To simplify the following equations, we consider the case in which the total orbital angular momentum of the electrons is L = 0; this is not an essential limitation. The K harmonics then depend on the two coordinates, α and θ, defined in §6. The element of solid angle in configuration space takes the form

$$d\hat{\Omega} = 8\pi^2 \sin^2\alpha \cos^2\alpha \sin\theta d\alpha d\theta \quad . \tag{12.6}$$

The operator Δ^* is of the form (6.4), so Eq. (12.2) can be solved by separation of variables. The solution of the more general case is described by Morse and Feshbach [49], who give the hyperspherical harmonics in terms of the coordinates α, θ_1, ϕ_1, θ_2, and ϕ_2, which determine the ratio of the distances and the directions of the two particles. Using their results, we obtain in our case

$$Y_{K\ell} = \frac{2^{\ell}\ell!\sqrt{(2\ell+1)(K/2+1)(K/2-\ell)!}}{\sqrt{\pi^3(K/2+\ell+1)!}} (\sin 2\alpha)^{\ell} C^{\ell}_{K/2-\ell}(\cos 2\alpha) \times$$
$$\times P_{\ell}(\cos\theta) \quad , \tag{12.7}$$

where

$$K = 0,2,4,\ldots \quad , \quad \ell = 0,1,\ldots,K/2 \quad , \tag{12.8}$$

C^n_m denotes a Gegenbauer polynomial, and P_ℓ is a Legendre polynomial. The quantum number ℓ has an immediate physical interpretation: It is the angular momentum of an individual electron (for L = 0 the angular momenta of both electrons are equal). The quantum number K has no immediate physical interpretation.

When the electrons are permuted, θ changes sign and α is replaced by $\pi/2-\alpha$. This implies that the K harmonic (12.7) is symmetric with respect to interchange of the electrons if $K/2-\ell$ is even and antisymmetric otherwise.

Simonov and Badalyan [42,43] used another set of hyperspherical harmonics, corresponding to a different choice of coordinates, in solving (12.2). They used coordinates A and λ defined by

$$A = \sqrt{\cos^2 2\alpha + \cos^2\theta \sin^2 2\alpha}\ , \quad \mathrm{tg}\lambda = \cos\theta\ \mathrm{tg}2\alpha \quad , \tag{12.9}$$

$$0 \le A \le 1 \quad , \quad 0 \le \lambda \le 2\pi \quad . \tag{12.10}$$

The operator Δ^* and the element of solid angle assume the following forms:

$$\Delta^* = 4\left[(1-A^2)\frac{\partial^2}{\partial A^2} + \left(\frac{1}{A} - 3A\right)\frac{\partial}{\partial A} + \frac{1}{A^2}\cdot\frac{\partial^2}{\partial\lambda^2}\right] \quad , \tag{12.11}$$

$$d\hat{\Omega} = \pi^2 A dA d\lambda \quad . \tag{12.12}$$

From (12.11) it is clear that the coordinates A and λ also allow a separation of variables in Eq. (12.2). The solutions are of the form

$$Y_{Kn} = \sqrt{\frac{K+2}{\pi^3}}\begin{pmatrix}\cos n\lambda \\ \sin |n|\lambda\end{pmatrix} A^{|n|} P^{(|n|,0)}_{(k/4)-(|n|/2)}(1-2A^2) \ . \tag{12.13}$$

For $n = 0$, the right side of Eq. (12.13) should in addition be divided by $\sqrt{2}$. For $n \ge 0$ in (12.13) $\cos n\lambda$ is used, and for $n < 0$ $\sin |n|\lambda$ is used. $P_\ell^{(n,m)}$ denotes a Jacobi polynomial. The quantum numbers K and n take the values

$$K = 0,2,4,\ldots \ , \quad n = -\frac{K}{2}, -\frac{K}{2}+2,\ldots,\frac{K}{2} \quad . \tag{12.14}$$

The harmonic (12.13) is symmetric with respect to interchange of the electrons if n is even and non-negative or odd and negative. It is convenient to use K harmonics of the form (12.13) for a system of three particles with identical masses, which occurs in nuclear physics problems.

§13. *Asymptotic Form for the Expansion in K Harmonics*

Let us consider the behavior of the expansion (12.5) in the asymptotic domain corresponding to ionization, i.e., for large r_1 and r_2. In this case we may ignore the quickly decaying terms from the discrete spectrum. Considering only the part of the expansion involving K harmonics, which is substituted into (6.3), we obtain, using (12.2) and (12.4), the system of radial equations

$$\left[\frac{d^2}{dr^2} - \frac{[K_i+(3/2)][K_i+(5/2)]}{r^2} + \kappa^2\right]F_i(r) = -\frac{2}{r}\sum_j Z_{ij}F_j(r) \ , \tag{13.1}$$

where

$$Z_{ij} = \int Y_i(\hat{\Omega})Z(\hat{\Omega})Y_j(\hat{\Omega})d\hat{\Omega} \quad . \tag{13.2}$$

In Eqs. (13.1) and (13.2) and later the one-dimensional indices i and j represent sets of quantum numbers specifying a K harmonic, and K_i denotes the corresponding value of K. In actual calculations only a finite number of equations in the system (13.1) can be taken into account. We denote this number by N. Then $i,j=1,2,\ldots,N$.

The system of equations (13.1) can be solved in terms of an asymptotic series. First the Coulomb interaction has to be diagonalized. Let Z_j and $\underset{\sim}{X}_j$ $(j=1,\ldots,N)$ be the eigenvalues and eigenvectors respectively of the matrix Z, which are defined by

$$\sum_{n=1}^{N} Z_{in} X_{nj} = Z_j X_{ij} \quad , \tag{13.3}$$

$$\sum_{n=1}^{N} X_{ni} X_{nj} = \sum_{n=1}^{N} X_{in} X_{jn} = \delta_{ij} \quad . \tag{13.4}$$

With the substitution

$$F_i = \sum_{j=1}^{N} X_{ij} f_j \quad , \tag{13.5}$$

we obtain, instead of (13.1),

$$\left[\frac{d^2}{dr^2} + \frac{2Z_i}{r} + \kappa^2\right] f_i = \frac{1}{r^2} \sum_{j=1}^{N} L_{ij} f_j \quad , \tag{13.6}$$

where

$$L_{ij} = \sum_{n=1}^{N} \left(K_n + \frac{3}{2}\right)\left(K_n + \frac{5}{2}\right) X_{ni} X_{nj} \quad . \tag{13.7}$$

Diagonalizing the Coulomb potential leads to a nondiagonal centrifugal potential, the matrix elements of which increase with increasing N.

The system of equations (13.6) has a solution in the form of an asymptotic series

$$f_j = \sum_{t=1}^{N} e^{i\rho + iW_t \ell n\, \rho} \sum_{n \geq 0} a_{tjn} \rho^{-n} \quad , \tag{13.8}$$

where

$$W_t = Z_t/\kappa \quad , \quad \rho = \kappa r \quad . \tag{13.9}$$

Substitution of (13.8) into (13.6) leads to the recursion relation

$$2(W_j - W_t - in)a_{tjn} = \sum_{m=1}^{N} L_{jm}a_{tm,n-1} +$$

$$+ [W_t^2 + i(2n-1)W_t - n(n-1)]a_{tj,n-1} \quad . \qquad (13.10)$$

From (13.10) it follows that the first coefficients of the series (13.8) are of the form

$$a_{tj0} = a_t\delta_{tj} \quad , \qquad (13.11)$$

where a_t can be chosen arbitrarily. The coefficients for $n > 0$ are uniquely determined by the recursion relation in terms of the preceding coefficients. Thus, the series (13.8) is completely determined once $a_1,\ldots,a_N$ are given.

Substituting (13.5) and (13.8) into (12.5) and ignoring the discrete states of the atom, we obtain

$$\Psi = \sum_{t=1}^{N} r^{-5/2} e^{i\rho + iW_t \ln\rho} \sum_{n\geq 0} \rho^{-n} \sum_{j=1}^{N} a_{tjn} y_j(\hat{\Omega}) \quad , \qquad (13.12)$$

where

$$y_j(\hat{\Omega}) = \sum_{i=1}^{N} X_{ij} Y_i(\hat{\Omega}) \quad . \qquad (13.13)$$

Because of the relations (13.4) the functions $y_1,\ldots,y_N$ are also orthonormal.

When (13.11) is taken into account, the first term of the asymptotic expansion (13.12) becomes

$$\Psi \sim \sum_{t=1}^{N} a_t y_t(\hat{\Omega}) r^{-5/2} e^{i\rho + iW_t \ln\rho} \quad . \qquad (13.14)$$

From the recursion relation (13.10) we see that the coefficients a_{tjn} with different values of the first index are defined independently. Therefore, each term of the sum over t in (13.12) is individually a solution of the Schrödinger equation (in the present approximation, where we consider a finite number of harmonics and ignore the discrete states of the atom).

The asymptotic expansion (13.12) differs from (6.5) in that it does not contain any powers of $\ell n\ \rho$. This difference can be explained as follows. According to (6.5) the wave function Ψ contains $\exp[iW(\hat{\Omega})\ \ell n\ \rho]$, so as a function of $\hat{\Omega}$ it oscillates more quickly as ρ is increased. This implies that in order to attain a given accuracy, it is necessary to include in the expansion (12.5) a larger number of harmonics as ρ increases. Therefore, the expansion (13.12) for a given number of harmonics N is valid only within a finite interval of ρ, where, on the one hand, ρ is large enough that the expansion (13.8) is a valid asymptotic solution of the system (13.6) and, on the other hand, it is not so large that N harmonics are insufficient to approximate the wave function. In contrast with this, the asymptotic expansion (6.5) has no upper limit on its range of validity. This difference may explain the absence of powers of $\ell n\ \rho$ in (13.12). In a finite interval and with a given finite error $\ell n\ \rho$ can be replaced by a finite series in inverse powers of ρ. Such a series is easy to obtain from the well-known expansion

$$\ell n\ \rho = \sum_{n=1}^{\infty} \frac{1}{n}\left(1 - \frac{1}{\rho}\right)^n \quad , \quad \rho > \frac{1}{2} \quad , \qquad (13.15)$$

if we retain only a finite number of terms.

The asymptotic forms (13.12) and (13.14) are uniquely determined once the coefficients $a_1,\ldots,a_N$ are given, which are thus analogs of the amplitude $A_{00}(\hat{\Omega})$. Let us establish the relationship between the coefficients a_t and $A_{00}(\hat{\Omega})$.

In our approximation the wave function is

$$\Psi = r^{-5/2} \sum_{t=1}^{N} F_t(r) Y_t(\hat{\Omega}) = r^{-5/2} \sum_{t=1}^{N} f_t(r) y_t(\hat{\Omega}) \ . \quad (13.16)$$

Note that because of the completeness of the system of K harmonics the series (13.16) can be fairly accurate at small distances as well if N is large enough. However, without explicitly considering the terms from the discrete spectrum of the atom, we expect the series to converge slowly, because the K harmonics do not reflect the specific nature of discrete-state excitation.

The coefficient a_t is determined from the asymptotic behavior of the function $f_t(r)$:

$$f_t \sim a_t e^{i\rho + iW_t \ln \rho} \ . \quad (13.17)$$

From the orthonormality of the functions $y_1, \ldots, y_N$ we obtain

$$f_t = r^{5/2} \int y_t(\hat{\Omega}) \Psi(r, \hat{\Omega}) d\hat{\Omega} \ . \quad (13.18)$$

Substitution of the asymptotic form (6.39) into (13.18) yields

$$f_t \sim e^{i\rho} \int y_t(\hat{\Omega}) A_{00}(\hat{\Omega}) e^{iW(\hat{\Omega}) \ln \rho} d\hat{\Omega} \ . \quad (13.19)$$

However, for $\rho \to \infty$ Eq. (13.19) does not reduce to (13.17) because the integral over $\hat{\Omega}$ falls off as $(\ln \rho)^{-1}$ owing to the oscillating factor $\exp[iW(\hat{\Omega}) \ln \rho]$. But, as already mentioned, the expansion (13.12), and consequently also Eqs. (13.16)-(13.19), make sense only when ρ is not too large. We choose ρ to be finite but large enough that (6.39) is valid, and we ascertain, for the chosen value of ρ, how the integral in (13.19) depends on N, the number of harmonics.

The dependence on N in (13.19) is determined by the function y_t. We see from (13.3) and (13.13) that for finite N the function $y_t(\hat{\Omega})$ is an approximate eigenfunction of an operator which multiplies by $Z(\hat{\Omega})$. In the limit $N \to \infty$ (since N determines the length of a one-dimensional segment, we should stipulate that we have in mind unlimited variation of all indices), the function y_t for some $\hat{\Omega} = \hat{\Omega}_t$ should have a δ function singularity, $\delta(\hat{\Omega} - \hat{\Omega}_t)$. We then have the relation

$$Z(\hat{\Omega})y_t(\hat{\Omega}) = Z(\hat{\Omega}_t)y_t(\hat{\Omega}) \quad . \tag{13.20}$$

The direction $\hat{\Omega}_t$ we shall call a characteristic direction, and the associated angles α and θ we shall call characteristic angles.

If $y_t(\hat{\Omega})$ has a δ-function dependence on $\hat{\Omega}$, from the completeness relation for the K harmonics

$$\sum_{i=1}^{\infty} Y_i(\hat{\Omega})Y_i(\hat{\Omega}_t) = \delta(\hat{\Omega} - \hat{\Omega}_t) \quad , \tag{13.21}$$

the coefficients in the expansion of y_t in K harmonics are seen to be proportional to $Y_i(\hat{\Omega}_t)$. Hence, allowing for (13.13) we should have for large N

$$X_{ij} \approx \frac{1}{g_j} Y_i(\hat{\Omega}_j) \quad , \tag{13.22}$$

where from (13.4) the normalizing denominator is

$$g_j^{\;2} = \sum_{i=1}^{N} [Y_i(\hat{\Omega}_j)]^2 \quad . \tag{13.23}$$

In addition, from (13.20) we have

$$Z_j \approx Z(\hat{\Omega}_j) \quad . \tag{13.24}$$

Indeed, substituting (13.22) and (13.24) into (13.3) and taking into account (13.2) and (13.21), we have

$$\sum_{n=1}^{\infty} Z_{in} Y_n(\hat{\Omega}_j) = \sum_{n=1}^{\infty} \int Y_i(\hat{\Omega}) Z(\hat{\Omega}) Y_n(\hat{\Omega}) Y_n(\hat{\Omega}_j) d\hat{\Omega}$$

$$= Z(\hat{\Omega}_j) Y_i(\hat{\Omega}_j) \quad . \qquad (13.25)$$

That (13.22) and (13.24) are satisfied has been verified by numerical calculation in the one-dimensional model discussed in §14. There $\hat{\Omega}_i$ is determined from (13.24) (with exact equality) and then it is verified that (13.22) is satisfied. We note that the eigenvalues of the matrix Z must lie within the range of variation of the function $Z(\hat{\Omega})$. Therefore, for any N there is at least one characteristic direction $\hat{\Omega}_j$ corresponding to each eigenvalue Z_j. In the multidimensional case $\hat{\Omega}_j$ is not determined uniquely from (13.24), and then one must consider also (13.22).

Equations (13.13) and (13.22) imply that

$$y_t(\hat{\Omega}) \approx \frac{1}{g_t} \sum_{j=1}^{N} Y_j(\hat{\Omega}_t) Y_j(\hat{\Omega}) \approx \frac{\delta(\hat{\Omega}-\hat{\Omega}_t)}{g_t} \quad . \qquad (13.26)$$

Substituting (13.26) into (13.19) and comparing the result with (13.17), we obtain

$$a_t = \frac{A_{00}(\hat{\Omega}_t)}{g_t} \quad . \qquad (13.27)$$

Equation (13.27) can also be obtained in a somewhat different manner from (13.14) by setting $\hat{\Omega}$ equal to one of the characteristic directions, $\hat{\Omega} = \hat{\Omega}_j$. Then the main contribution to the sum over t comes from the term with $t = j$. From (13.13), (13.22), and (13.23) we have

$$y_j(\hat{\Omega}_j) \approx g_j \quad , \tag{13.28}$$

which leads to

$$\Psi(r,\hat{\Omega}_j) \sim a_j g_j r^{-5/2} e^{i\rho+iW(\hat{\Omega}_j)\,\ln\rho} \quad . \tag{13.29}$$

Comparison of this expression with (6.39) corroborates Eq. (13.27).

The meaning of relation (13.27) is quite transparent. The coefficient a_t describes the scattering in the characteristic direction $\hat{\Omega}_t$. The δ-function dependence of the functions $y_1,\ldots,y_N$ on $\hat{\Omega}$ assigns to each direction $\hat{\Omega}_t$ the appropriate charge $Z(\hat{\Omega}_t)$ and ensures the correct logarithmic phase shift.

Equation (13.27) is inconvenient for practical application in that it is difficult to determine $\hat{\Omega}_t$. However, this relation can be transformed to a more convenient form by employing the expansion of the amplitude $A_{00}(\hat{\Omega})$ in K harmonics. Let

$$A_{00}(\hat{\Omega}) = \sum_{m=1}^{M} c_m Y_m(\hat{\Omega}) \quad . \tag{13.30}$$

Substituting (13.30) into (13.27) and using (13.22), we find

$$a_t = \sum_{m=1}^{M} c_m X_{mt} \quad . \tag{13.31}$$

This result has the same degree of accuracy as (13.27), since (13.22) was used in the derivation of both of these expressions.

Equation (13.31) is equivalent to

$$a_t = \int y_t(\hat{\Omega}) A_{00}(\hat{\Omega})\, d\hat{\Omega} \quad . \tag{13.32}$$

It is necessary that all harmonics in (13.30) should be included among the N harmonics being considered. The correctness of (13.32) can also be demonstrated by substituting into it

Eq. (13.26) and comparing the result obtained with (13.27). From (13.22) it follows that

$$A_{00}(\hat{\Omega}) = \sum_{t=1}^{N} a_t y_t(\hat{\Omega}) \quad . \tag{13.33}$$

Equation (13.32) is rigorously valid when $Z(\hat{\Omega}) = \text{const}$. Then we can set $X_{mt} = \delta_{mt}$, i.e., $y_t = Y_t$, and (13.32) reduces to $a_t = c_t$. Because of the presence of an $\hat{\Omega}$-dependent function $Z(\hat{\Omega})$ in the exponential in Eq. (6.39), the wave function depends on $\hat{\Omega}$ in a more complicated way than in (13.30). Therefore, in order that the series (13.16) approximate the wave function well, the number of harmonics in the expansion of the wave function should be considerably greater than the number of harmonics in the expansion of the amplitude, i.e., we require that $N \gg M$. When $Z(\hat{\Omega}) = \text{const}$, it is sufficient that $N = M$; moreover, the system (13.1) splits up into independent equations. Thus, use of the direction-dependent Coulomb potential requires considerably more computational work.

§14. *Model Calculation*

In order to get an idea of the limits of validity of the asymptotic expansions (6.5) and (13.12), and of their consistency, Peterkop and Rabik [28] calculated numerical results for a simplified model. The function $Z(\hat{\Omega})$ defined in (6.2) was replaced by

$$Z(\alpha) \equiv Z(\alpha,\pi) = \frac{1}{\cos\alpha} + \frac{1}{\sin\alpha} - \frac{1}{\sin\alpha + \cos\alpha} \quad , \tag{14.1}$$

which corresponds to replacing (1.2) by

$$V = -\frac{1}{r_1} - \frac{1}{r_2} + \frac{1}{r_1 + r_2} \quad . \tag{14.2}$$

This potential energy corresponds to the case in which both electrons move along one line in different directions from the nucleus. According to Wannier [50] this situation occurs in the ionization of an atom by an electron if the energy is close to threshold (see §24). However, as $E \to 0$ the domain of validity of the asymptotic expansions quickly recedes; therefore, the calculations in [28] were performed for comparatively high energies.

Note that $Z(\alpha)$ is symmetric under the substitution $\alpha \to \pi/2-\alpha$, has a minimum at $\alpha = \pi/4$, increases monotonically with increasing distance from this point, and goes to infinity for $\alpha = 0$ and $\alpha = \pi/2$.

The wave function Ψ used in the model [28] was also assumed to be independent of the angle θ. The expansion in K harmonics using the form (12.7) then reduces to an expansion in a Fourier series in the functions

$$Y_i(\alpha) \equiv Y_{K0}(\hat{\Omega}) = \frac{\sin(K+2)\alpha}{\pi^{3/2} \sin 2\alpha} \quad . \tag{14.3}$$

The element of solid angle takes the form

$$d\hat{\Omega} = 16\pi^2 \sin^2 \alpha \cos^2 \alpha d\alpha \quad . \tag{14.4}$$

The coefficients of the asymptotic expansion (6.5) were calculated by means of the recursion relation (6.6). The amplitude A_{00} was chosen of the form

$$A_{00}(\alpha) = \frac{\sin(K+2)\alpha}{\sin 2\alpha} \quad . \tag{14.5}$$

The calculations were performed for K = 0 and K = 16. In order to simplify the recursion relation, Peterkop and Rabik made the transformation

$$A_{nm}(\alpha) = \frac{a_{nm}(\alpha)}{\sin 2\alpha} \quad , \tag{14.6}$$

which brings the operators D_1 and D_2 to the form

$$D_1 = 2\frac{dW}{d\alpha}\cdot\frac{d}{d\alpha} + \frac{d^2W}{d\alpha^2}\ , \quad D_2 = \frac{d^2}{d\alpha^2} + \frac{1}{4} - W^2\ . \tag{14.7}$$

The main difficulty in using the recursion relation (6.6) arises from the calculation of the derivatives. To determine a_{nm} for $n \le n_{max}$, it is necessary to know the derivatives of $a_{00}(\alpha)$ and $Z(\alpha)$ up to order $2n_{max}$. The derivatives of $a_{00}(\alpha)$ are easily calculated. To find the derivatives of $Z(\alpha)$, Peterkop and Rabik obtained the equation

$$\frac{d^nZ}{d\alpha^n} = \sum_{j=1}^{n+1} d_{nj}\left[\frac{(\sin\alpha)^{j-1}}{(\cos\alpha)^j} + \frac{(-\cos\alpha)^{j-1}}{(\sin\alpha)^j} - \frac{(\sin\alpha - \cos\alpha)^{j-1}}{(\sin\alpha + \cos\alpha)^j}\right], \tag{14.8}$$

where $d_{0j} = \delta_{1j}$, and the remaining coefficients are defined by the recursion relation

$$d_{n+1,j} = (j-1)d_{n,j-1} + jd_{n,j+1}\ . \tag{14.9}$$

Then the coefficients a_{1m} $(m = 0,1,2)$ and their derivatives up to order $2(n_{max} - 1)$ were found by means of Eq. (6.6) and the equations resulting from repeated differentiation of Eq. (6.6). It was then possible to find the next coefficients a_{2m} $(m = 0, 1,2,3,4)$ and their derivatives up to order $2(n_{max} - 2)$, and so on. Derivatives of products were determined by Leibnitz' formula. The computations were performed on a GE-415 computer, programmed in FORTRAN. The coefficients a_{nm} were calculated for all possible values of m with $n \le n_{max}$ for various values of κ and α. Then the partial sums of the series (6.5), taking into account all the terms with $n \le 1,2,\ldots,n_{max}$, were determined for different values of r. The computer program allowed the value of n_{max} to be varied. The majority of the calculations were performed with $n_{max} = 8$.

We note some of the properties of the coefficients a_{nm} as functions of n, m, κ, and α. The behavior of the coefficients with the maximum index m is given by Eqs.(8.2) and (8.5). The behavior of the rest of the coefficients is more complicated. For a given value of n the coefficients oscillate as m changes and attain their maximum absolute values for small or intermediate values of m. For a specified m the coefficients also oscillate as n changes, but their absolute values grow rapidly with increasing n. This growth is all the more rapid the smaller κ becomes or the closer α is to 0 (or π/2). For example, when K = 0 (i.e., $a_{00} = \sin 2\alpha$) and $\alpha = \pi/4$ the orders of magnitude of the coefficient $a_{5,0}$ for κ = 0.5, 1, and 2 are 10^7, 10^6, and 10^5, respectively; for $\alpha = \pi/20$ they are 10^{17}, 10^{15}, and 10^{13}. At α = 0 (or π/2) the function Z(α) and all the coefficients a_{nm} with n > 0 go to infinity.

As noted, the series (6.5) is to be interpreted in the asymptotic sense because the sum over n diverges. In practice, however, the series may be considered to be convergent if the component sums over m, as functions of n, show well-defined deep minima over some interval. The series can then be restricted to this interval. In this event we shall say that the series converges. For r sufficiently large, there is always an interval containing a minimum. The limiting values of r, i.e., those for which there are no intervals with minima, are determined numerically. The results of the calculations are shown in Fig. 2 (r is shown on a logarithmic scale). Since the amplitude (14.5) is symmetric or antisymmetric under the transformation $\alpha \to \pi/2-\alpha$, the wave function Ψ has the same property. Therefore, it suffices to consider α in the range $0 \le \alpha \le \pi/4$. Note that the values K = 0 and K = 16 used in the calculations correspond to the symmetric case.

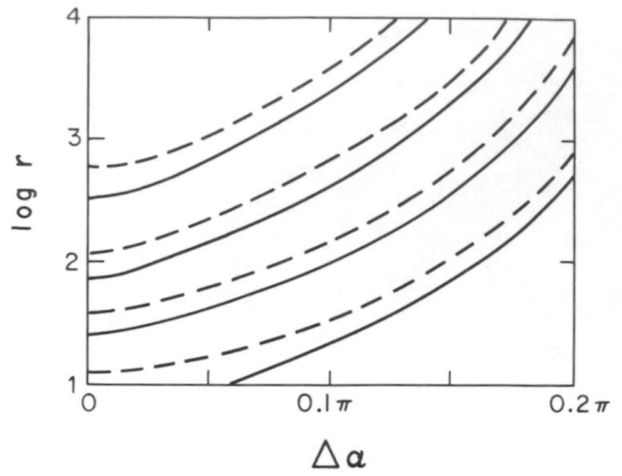

Figure 2. Limits of validity of expansion (6.5). The solid curves are for $a_{00} = \sin 2\alpha$; the dashed curves for $a_{00} = \sin 18\alpha$; the numbers indicate the value of κ; $\Delta\alpha = |\alpha - \pi/4|$.

We see from Fig. 2 that the limits of validity of expansion (6.5) increase rapidly as α departs from $\pi/4$. For example, when $K = 0$ and $\kappa = 1$, the limit of validity for $\alpha = \pi/4$ is at $r \approx 80$ and for $\alpha = \pi/10$ it is at $r \approx 2000$, an increase by a factor of 25. At the same time the function $Z(\alpha)$ at these points is equal to 2.1 and 3.5, an increase by a factor of only 1.7. Thus, Eq. (10.6) is not sufficient for estimating the limits of validity because of the large role played by the derivatives of $Z(\alpha)$, which grow considerably faster than $Z(\alpha)$ as the difference between α and $\pi/4$ increases. The domain of validity also recedes if the energy $E = \kappa^2/2$ is decreased. For example, for $K = 0$, $\kappa = 0.5$, and $\alpha = \pi/4$ the limit of validity is at $r \approx 300$. This limit also increases if the amplitude A_{00} is a harmonic of higher order, but the increase is not very great, which can also be explained by the importance of the derivatives of $Z(\alpha)$. It was found that at $\alpha = \pi/4$ for n large and even

$$\frac{d^n Z}{d\alpha^n} \sim 2.55 \left(\frac{4}{\pi}\right)^n n! \qquad (14.10)$$

The derivatives of odd order are equal to zero at $\alpha = \pi/4$. For $\alpha \neq \pi/4$ the even derivatives grow faster with increasing order than Eq. (14.10) indicates.

Direct numerical investigation of a divergent asymptotic series cannot be considered a completely rigorous procedure. Erroneous cases of apparent convergence in the calculation of Peterkop and Rabik would appear to be exceptional, because results for different r and α were considered and also because the results agree with those obtained from the series (13.12). The series (13.12) was checked, in addition, by numerically integrating the system of radial equations (13.1).

The series (13.12) was evaluated on the GE-415 computer. The coefficients a_L were determined from Eq. (13.31). Only one choice of amplitude was employed, namely, $a_{00} = \sin 2\alpha$. Equation (13.31) then gives $a_t = \pi^{3/2} X_{1t}$. The fact that the selected amplitude contains only one harmonic is not very important, because the solution for an amplitude in the form of the sum (13.30) is equal to the sum of the solutions for the individual terms in (13.30). This is true also for the evaluation of the series (6.5). More important is the fact that the amplitude was chosen to be the lowest harmonic. As the amplitude oscillates more rapidly, and considering that it is multiplied by $\exp[iW(\hat{\Omega}) \ln \rho]$, we expect that a larger number of harmonics will have to be included in the expansion of the wave function. The choice of the amplitude in the form $\sin 2\alpha$ corresponds to the symmetric case with respect to permutation of the electrons; therefore, only those harmonics (14.3) were taken into account for which $K = 4j$, where $j = 0,1,2,\ldots$ The calculations were performed with different numbers of harmonics: $N = 5,10,20,30,40$. These correspond to maximum values $K_{max} = 16$, 36, 76, 116, 156. The main calculations were carried out for $N = 20$ ($K_{max} = 76$).

Computations were also made to check whether (13.22) and (13.24) are valid. In the one-dimensional model, Eq. (13.24) uniquely determines the characteristic angle α_j. The transcendental equation (13.24) was solved numerically on the computer by the method of bisection of intervals. It turned out that the characteristic angles were distributed quite uniformly. For example, for $N = 5$ the angles are 0.10, 0.25, 0.40, 0.56, and 0.71 rad. Owing to the symmetry of $Z(\alpha)$ the angle $\pi/2 - \alpha_j$ is also a characteristic angle. With the characteristic angles which were computed from (13.24) the absolute value of the difference between the left and right sides of (13.22), averaged over i and j, was found to be 0.0091, 0.0058, and 0.0032, respectively, for $N = 5$, 10, and 20. As N increases, the quantities X_{ij} fall off as $N^{-1/2}$ by virtue of being normalized to unity. However, the mean differences given above fall off faster. Thus, with increasing N Eqs. (13.22) and (13.24) become more accurate.

We note that in the one-dimensional case we can find an explicit expression for the sum

$$\sum_{i=1}^{N} Y_i(\alpha)Y_i(\alpha_j) = \frac{1}{4\pi^3 \sin 2\alpha \sin 2\alpha_j}\left[\frac{\sin 4N(\alpha-\alpha_j)}{\sin 2(\alpha-\alpha_j)} - \frac{\sin 4N(\alpha+\alpha_j)}{\sin 2(\alpha+\alpha_j)}\right]. \tag{14.11}$$

For $N \to \infty$ we obtain

$$\sum_{i=1}^{N} Y_i(\alpha)Y_i(\alpha_j) \sim \frac{\delta(\alpha-\alpha_j) + \delta(\alpha+\alpha_j-\pi/2)}{8\pi^2 \sin 2\alpha \sin 2\alpha_j}, \tag{14.12}$$

$$g_j \sim \frac{1}{\pi \sin 2\alpha_j} \sqrt{\frac{N}{2\pi}} \quad . \tag{14.13}$$

From (13.21) and (14.4) it follows that

$$\sum_{i=1}^{\infty} Y_i(\alpha)Y_i(\alpha_j) = \frac{\delta(\alpha - \alpha_j)}{4\pi^2 \sin 2\alpha \sin 2\alpha_j} \quad . \tag{14.14}$$

The difference between (14.12) and (14.14) is explained by the fact only symmetric harmonics are considered in (14.12).

In Figs. 3 and 4 we compare[1)] the expansions (6.5) and (13.12), calculated with $Z(\alpha)$ of the form (14.1). The quantity $\Delta\alpha = |\alpha - \pi/4|$ is taken as the independent variable. From Fig. 3, in which the first terms of the various series are compared, we see that the series (13.14) approaches (6.39) as N increases. This corroborates the validity of Eq. (13.31). The difference between (13.14) and (6.39) becomes larger as $\Delta\alpha$ increases. This is to be expected, because as $\Delta\alpha$ increases, $Z(\alpha)$ also increases, $\exp\{i[Z(\alpha)/\kappa] \ln \rho\}$ oscillates more rapidly, and for fixed N the series (13.14) becomes less accurate. From Fig. 4, which compares the complete sums (up to the interval containing a minimum) of the series (6.5) and (13.12), it is also seen that the series are in good agreement for small $\Delta\alpha$. The examples given here, and also the results of calculations for other cases, indicate that when both expansions (6.5) and (13.12) converge (in the asymptotic sense) and when N is sufficiently large, their sums are the same.

The lower limits of validity of the asymptotic expansion (13.12) correspond roughly to those given in Fig. 2. Unlike (6.5), the expansion (13.12) also has upper limits of validity.

[1)] Figures 3-5 and Tables 1 and 2 were not given in [28].

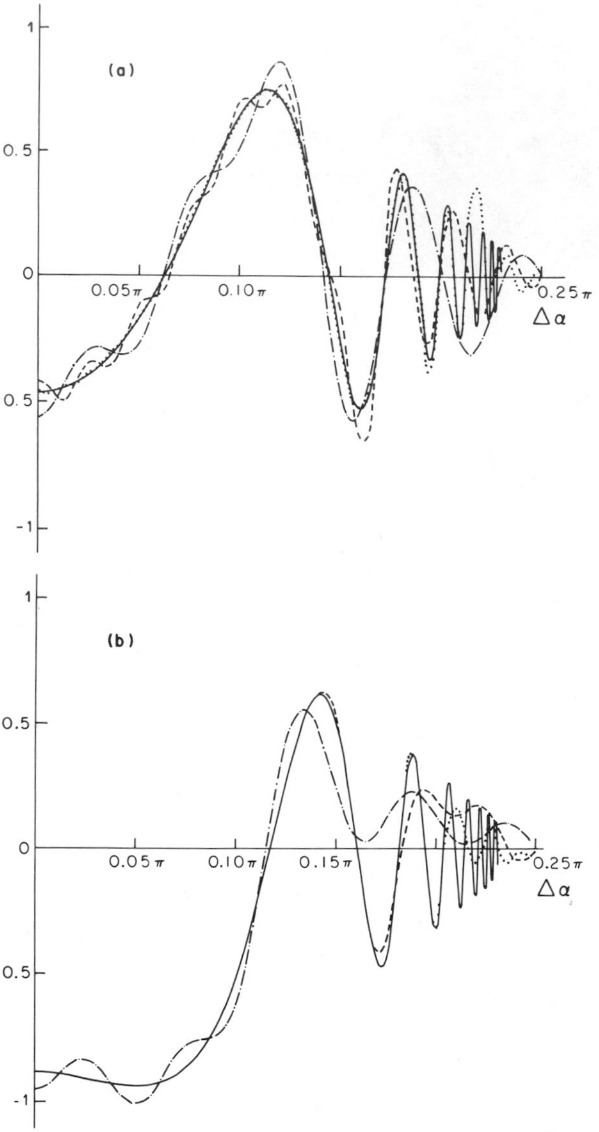

Figure 3. First terms of the asymptotic series, multiplied by $r^{5/2}\sin 2\alpha$, with $a_{00} = \sin 2\alpha$, $\kappa = 2$, $r = 200$: a) real part; b) imaginary part; $\Delta\alpha = |\alpha-\pi/4|$; —— from Eq. (6.39); -·-·-· from Eq. (13.34) with N = 10; — — — with N = 20; · · · with N = 40.

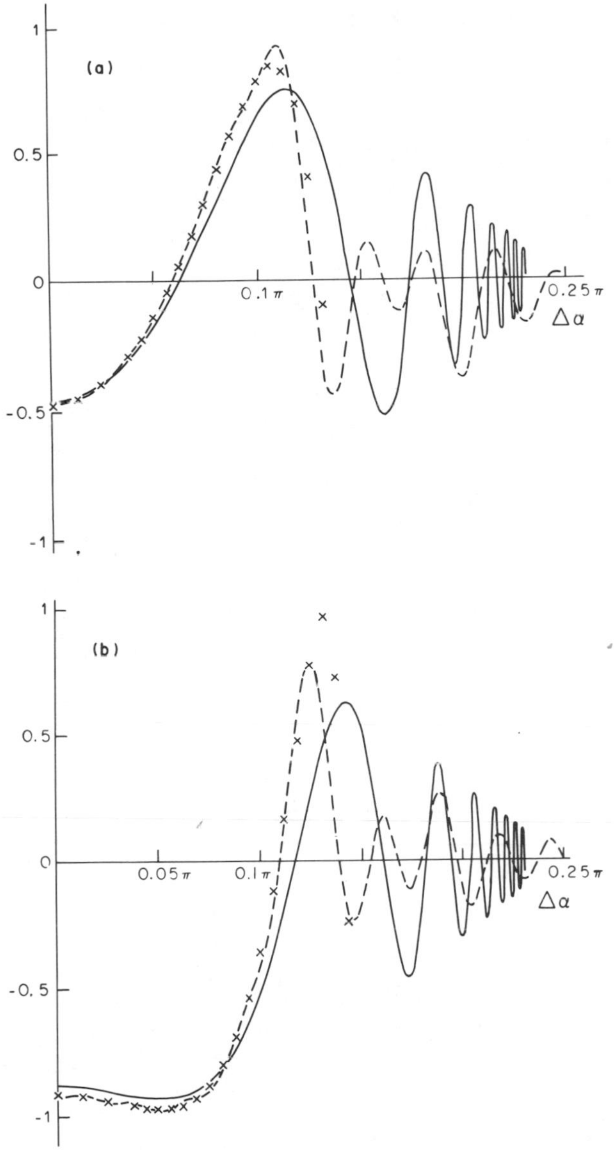

Figure 4. Sums of the series (6.5) and (13.12), multiplied by $r^{5/2} \sin 2\alpha$, with $a_{00} = \sin 2\alpha$, $\kappa = 2$, $r = 200$: a) real part; b) imaginary part; $\Delta\alpha = |\alpha - \pi/4|$; × from Eq. (6.5); — — — from Eq. (13.12) with N = 20; —— first term of (6.5).

As N increases, the upper limits increase, but along with this the lower limits for applying the expansion (13.8) to the system of equations (13.6) also increase. The principal reason for this is that the matrix elements L_{ij} grow rapidly as N increases. These matrix elements can be approximated using (13.7), (13.22) and the second derivative with respect to α of the approximate δ function in the form (4.11). For the diagonal elements we obtain

$$L_{ii} \approx \frac{16}{3} N^2 \quad . \tag{14.15}$$

The mean values of the diagonal elements of the centrifugal matrix are found to be 132, 532, and 2132, respectively, for N = 5, 10, and 20. These values are only about one unit less than the numbers obtained from (14.15). The off-diagonal matrix elements grow more slowly, and are apparently approximately proportional to N. Their mean absolute values are 41, 90, and 193.

In addition, as N increases, characteristic angles closer to 0 (or $\pi/2$) are found and, accordingly, large eigenvalues $Z_t = Z(\alpha_t)$ are obtained. From the recursion relation (13.10) we see that as L_{jm} and Z_t increase the coefficients a_{tjn} with $n > 0$ and, hence, the limits of validity of (13.8) also become larger.

If the expansion (6.5) converges (in the asymptotic sense) for given r at one value of α, then it could still diverge for some other α farther from $\pi/4$. In contrast with this, the convergence of the series (13.8), and in practice also (13.12), does not depend on α. Figure 4 shows the result of summing the series (13.12) for all values of α, but this series can be considered correct only in the limited region where it coincides with (6.5). As α departs farther from $\pi/4$ the function $Z(\alpha)$ grows without bound, the wave function oscillates more rapidly, and in analogy to the case of increasing r, ever larger values of N are required for a good approximation. For a given N the series (13.12) can be applied only in some interval $\alpha_{min} \le \alpha \le \pi/2-\alpha_{min}$. As N increases, α_{min} decreases. It is clear from Fig. 3 that as N increases the interval where (13.14) agrees with the given exact value (6.39) is extended. Agreement of the first terms should be considered a necessary condition for the validity of (13.12). As N increases, so does the lower limit of validity of (13.12). As a result these lower limits are similar in shape to the curves given in Fig. 2; we may conclude from the calculated data that these lower limits agree roughly with the curves of Fig. 2 also in absolute value.

The unbounded growth of the function $Z(\alpha)$ as $\alpha \to 0$ (or $\pi/2$) makes the Fourier series converge slowly. The presence of singularities is an additional difficulty in the real Coulomb problem (the primary difficulty is the fact that Z depends on α). Therefore, a comparison of the asymptotic expansions (6.5) and (13.12) for a nonsingular function $Z(\alpha)$ is of some interest. In [28] we also performed the calculations with the nonsingular potential obtained by replacing (14.1) by the first terms in the expansion in powers of $\alpha-\pi/4$:

$$Z(\alpha) \to \frac{3}{\sqrt{2}} + \frac{11}{2\sqrt{2}}\left(\alpha - \frac{\pi}{4}\right)^2 . \tag{14.16}$$

Wannier [50] showed that the function (14.16) has physical significance at low energies (see §24). Its coefficients determine the threshold behavior of the ionization cross section.

If (14.16) is used, the coefficients of the series (6.5) and (13.12) grow considerably more slowly with increasing n, and the limits of validity are pushed in the direction of smaller r. In addition, they depend much less strongly on α. Figure 5 shows that the series (13.12) for N = 20 agrees well

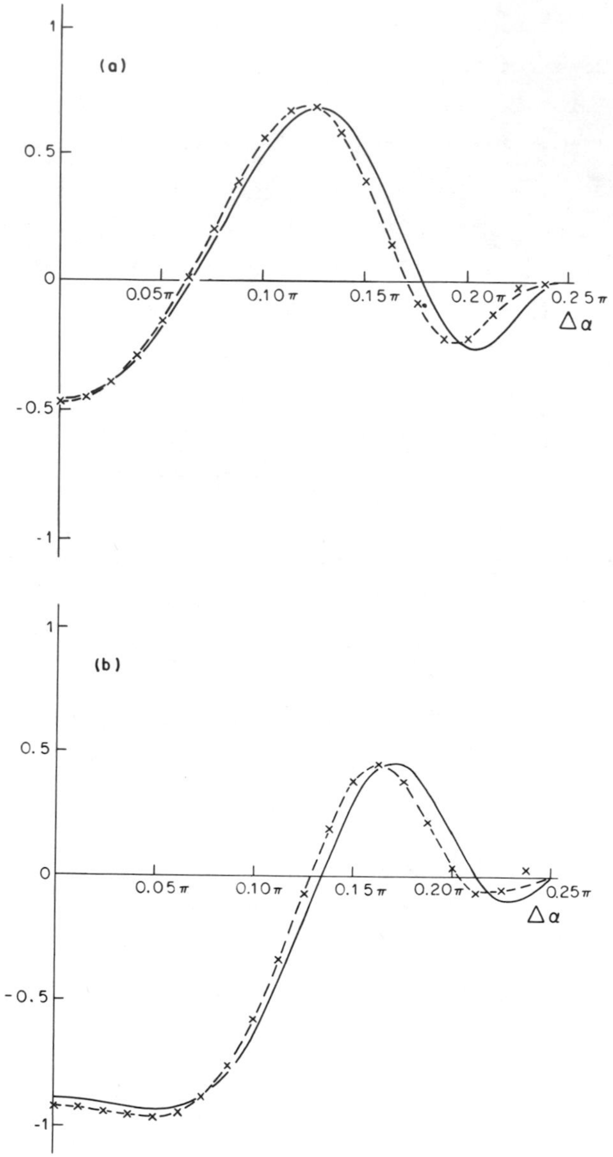

Figure 5. Sums of the series (6.5) and (13.12) with the function (14.16). The parameters and notation are the same as in Fig. 4.

with (6.5) for almost all α. The results for the two series agree in many cases to three or four digits. Table 1 gives, as an example, a comparison of the convergence of the two series. Clearly, the first terms ($n_{max} = 0$) of the two series are the same. The sums of the first two terms are also similar, but when the later terms are taken into account, the series (6.5) quickly approaches a stationary value, while the series (13.12) begins to oscillate. Only for $n_{max} \approx 15$ does it approach a stationary value, which nevertheless agrees well with (6.5). A similar situation, but with less precise agreement of the final results, was found also in calculations using Z in the form (14.1). The slower convergence of the series (13.12) is apparently explained by the absence of powers of logarithms. If we consider that for each n (6.5) contains 2n+1 terms of the sum over m, then the overall number of terms in (6.5) for $n \le 3$ is 16, which is equal to the number of terms

Table 1

Comparison of the convergence of the series (6.5) and (13.12) (multiplied by $r^{5/2} \sin 2\alpha$) when (14.16) is used with $a_{00} = \sin 2\alpha$, $\kappa = 2$, $r = 200$, $\alpha = \pi/10$

n_{max}	(6.5)	(13.12), N = 20
0	0.5056+i0.2997	0.5054+i0.2998
1	0.4123+i0.3926	0.4122+i0.3907
2	0.3947+i0.3845	0.4027+i0.3850
4	0.3944+i0.3815	0.3495+i0.3764
6	0.3944+i0.3815	0.4799+i0.3872
8	0.3944+i0.3815	0.3177+i0.3785
10	--	0.4359+i0.3834
12	--	0.3802+i0.3806
15	--	0.3945+i0.3832
20	--	0.3947+i0.3816
25	--	0.3947+i0.3816
30	--	0.3947+i0.3816

in the sum over n in (13.12) that are important in practice. The data of Table 1 were calculated for $\alpha = \pi/10$. For values of α closer to $\pi/4$ the sums of the series (6.5) and (13.12) agree even better. With $\alpha = \pi/4$ and the rest of the parameters equal to those given in Table 1, the series (6.5) attains the stationary value -0.47309-i0.92015 for $n_{max} = 4$, and the series (13.12) attains the stationary value -0.47305-i0.92016 for $n_{max} = 22$; these results differ only in the fifth significant digit.

In order to check the correctness of the results from the asymptotic expansion, we have also integrated numerically the system of radial equations. The initial values for some large r were determined with the aid of the asymptotic expansion (13.8), and then the radial functions for smaller values of r were found by numerical integration. The initial system of equations (13.1) is more convenient for numerical integration, because it does not contain the large matrix elements L_{ij}. The transformation to this system is carried out with the aid of (13.5). The results indicate that if the asymptotic expansion is applicable for the smaller values of r, then it agrees with the results of numerical integration. In Table 2 we compare the series (6.5) and (13.12) with the results of numerical

Table 2

Comparison of the series (6.5) and (13.12) with the results of numerical integration. All parameters except r correspond to those in Table 1

	r = 150	r = 100
(6.5)	0.3196+i0.4294	0.2851+i0.4184
(13.12), N = 20	0.3189+i0.4297	--
Num. Integ., N = 20	0.3187+i0.4299	0.2884+i0.4181

integration. In (13.1) we consider N = 20 equations. The initial values for the integration were determined with the aid of (13.8) with r = 200. The numerical integration was performed by the method of de Vogelaere [see 51] with step size 0.1. The example in Table 2 shows good agreement among the results. There is no value given for the series (13.12) for r = 100 because the series does not converge for N = 20. The result of the series (6.5) should be considered the most accurate, since the errors accumulate in the numerical integration and the series (13.12) contains some error due to N being finite. The agreement among these results confirms the correctness of the asymptotic expansions (6.5) and (13.12) and of Eq. (13.31), which is associated with them.

The asymptotic expansions (13.8) and (13.12) are applied in practice when the ionization problem is solved by numerically integrating the system of equations following from the expansion (12.5) (with the discrete levels included). This system is usually integrated outwards from r = 0. At a sufficiently large value of r the numerical solution (or more precisely, a linear combination of independent solutions) is matched with expansion (13.8). The matching conditions determine the coefficients a_t, which in turn determine the ionization amplitude according to (13.33).

CHAPTER IV

MULTIDIMENSIONAL COULOMB WAVE FUNCTIONS

The problem of many charged particles is described by Eq. (7.27), which can be viewed as an equation for a single particle moving in a multidimensional configuration space in the field of a direction-dependent Coulomb potential. Equation (7.27) cannot be solved in general form. It is useful therefore to consider the simpler problem, in which Z does not depend on $\hat{\Omega}$. We can then find exact analytical solutions, which prove useful in studying the properties of solutions of the more complicated problem. The asymptotic form of the Coulomb wave function when Z = const can be determined exactly and used to check the general equations (9.1) and (9.2) in the limit $Z(\hat{\Omega}) \to \text{const}$. The multidimensional Coulomb wave functions are also used in constructing integral expressions for the ionization amplitude (see §§17 and 18).

§15. *Plane Wave in Multidimensional Space*

Before treating the Coulomb case, let us note some of the properties of a plane wave in multidimensional space. The plane-wave results are used directly in determining the asymptotic behavior of certain integals; it is also interesting to compare them with the corresponding equations in the Coulomb case.

An n-dimensional plane wave is an immediate generalization of a three-dimensional plane wave

$$\psi_n^{(0)}(\underset{\sim}{k},\underset{\sim}{r}) = \exp(i\underset{\sim}{k}\cdot\underset{\sim}{r}) \quad , \qquad (15.1)$$

where $\underset{\sim}{k}$ and $\underset{\sim}{r}$ are n-dimensional vectors;

$$\underset{\sim}{k}\cdot\underset{\sim}{r} = k_1x_1 + \ldots + k_nx_n \quad . \qquad (15.2)$$

The plane wave $\psi_n^{(0)}$ satisfies the equation

$$(\Delta + \kappa^2)\psi = 0 \quad , \qquad (15.3)$$

where

$$\Delta = \sum_{j=1}^{n} \frac{\partial^2}{\partial x_j^{\,2}} \quad , \quad \kappa^2 = \sum_{j=1}^{n} k_j^{\,2} \quad . \qquad (15.4)$$

In hyperspherical coordinates (7.1)-(7.4) with the polar axis taken along the vector $\underset{\sim}{k}$, Eq. (15.1) takes the form

$$\psi_n^{(0)} = \exp(i\kappa r\cos\theta_1) \quad . \qquad (15.5)$$

In certain applications it is necessary to know the asymptotic behavior as $r \to \infty$ of integrals like

$$I = \int F(\hat{\Omega})\exp(i\underset{\sim}{k}\cdot\underset{\sim}{r})d\hat{\Omega} \quad , \qquad (15.6)$$

where $F(\hat{\Omega})$ is a sufficiently smooth function.

Let us substitute (7.9) into (15.6) and integrate first over θ_1. If n is odd, we integrate (n-1)/2 times by parts; if n is even, we integrate (n/2)-1 times by parts, and then we apply Eqs. (11.14)-(11.17). We obtain as a result

$$\int_0^{\pi} F(\theta_1)\exp(i\kappa r\cos\theta_1)(\sin\theta_1)^{n-2}d\theta_1 \sim$$

$$\sim \frac{1}{2}\,\Gamma(\lambda)\left[F(0)\left(\frac{2}{i\kappa r}\right)^{\lambda} e^{i\kappa r} + F(\pi)\left(\frac{2i}{\kappa r}\right)^{\lambda} e^{-i\kappa r}\right], \qquad (15.7)$$

where

$$\lambda = \frac{n-1}{2} \quad . \tag{15.8}$$

For the complex numbers raised to powers we take the principal value of the argument.

We see from (7.1)-(7.4) that when θ_1 is equal to zero or π, all values of the remaining angles correspond to the same point on the sphere. Hence, if $F(\hat{\Omega})$ is continuous on the sphere, then when θ_1 is equal to θ or π, the function $F(\hat{\Omega})$ does not depend on the magnitudes of the remaining angles. Integrating over the remaining angles gives a factor equal to the surface area of a sphere in n-1 dimensional space. Using (7.10) we obtain

$$I \sim F(\hat{\Omega}_{\underset{\sim}{k}}) \left(\frac{2\pi}{i\kappa r}\right)^{\lambda} e^{i\kappa r} + F(-\hat{\Omega}_{\underset{\sim}{k}}) \left(\frac{2\pi i}{\kappa r}\right) e^{-i\kappa r} \quad , \tag{15.9}$$

which for n = 3 agrees with the result quoted in §124 of Landau and Lifshitz [8]. We note that an equivalent form of (15.9) is

$$e^{i\underset{\sim}{k}\cdot\underset{\sim}{r}} \sim \delta(\hat{\Omega}_{\underset{\sim}{k}} - \hat{\Omega}_{\underset{\sim}{r}}) \left(\frac{2\pi}{i\kappa r}\right)^{\lambda} e^{i\kappa r} + \delta(\hat{\Omega}_{\underset{\sim}{k}} + \hat{\Omega}_{\underset{\sim}{r}}) \left(\frac{2\pi i}{\kappa r}\right)^{\lambda} e^{-i\kappa r} \, . \tag{15.10}$$

The stationary points correspond to the cases in which the vectors $\underset{\sim}{k}$ and $\underset{\sim}{r}$ are parallel or antiparallel. Equations equivalent to (15.9) have been obtained by Lieber, Rosenberg, and Spruch [52] and by Peterkop [53].

Now let us look at the expansion of a plane wave in partial waves. For this we can apply the equation [26]

$$e^{i\kappa r\cos\theta} = \Gamma(\mu)\left(\frac{2}{\kappa r}\right)^{\mu} \sum_{K=0}^{\infty} i^{K}(K+\mu) J_{K+\mu}(\kappa r) C_K^{\mu}(\cos\theta) \quad , \tag{15.11}$$

where $\mu > 0$, $J_{K+\mu}$ is a Bessel function, and C_K^{μ} is a Gegenbauer polynomial. It is convenient to choose μ such that each term of the sum (15.11) individually satisfies Eq. (15.3). This occurs for

$$\mu = \frac{n}{2} - 1 = \lambda - \frac{1}{2} \quad . \tag{15.12}$$

Henceforth we shall consider μ to be defined by (15.12).

We introduce the free-particle radial wave function

$$R^{(0)}_{\kappa K}(r) = r^{-\lambda}\sqrt{\kappa r}\, J_{K+\mu}(\kappa r) \quad . \tag{15.13}$$

Using the asymptotic form of the Bessel function, we have

$$R^{(0)}_{\kappa K} \sim \sqrt{\frac{2}{\pi}}\, r^{-\lambda} \sin\left[\kappa r - (K+\lambda-1)\frac{\pi}{2}\right] \quad . \tag{15.14}$$

The radial function is normalized so that

$$\int_0^\infty R^{(0)}_{\kappa K}(r) R^{(0)}_{\kappa' K}(r) r^{n-1} dr = \delta(\kappa - \kappa') \quad . \tag{15.15}$$

Utilizing the connection between the Bessel function and the confluent hypergeometric function [26], we can rewrite Eq. (15.13) in the form

$$R^{(0)}_{\kappa K} = A^{(0)}_{\kappa K}(2\kappa r)^K e^{-i\kappa r}\, {}_1F_1(K+\lambda, 2K+2\lambda, 2i\kappa r) \quad , \tag{15.16}$$

where

$$A^{(0)}_{\kappa K} = \frac{(2\kappa)^\lambda}{\sqrt{2\pi}} \cdot \frac{\Gamma(K+\lambda)}{\Gamma(2K+2\lambda)} \quad . \tag{15.17}$$

The result (15.16) is similar in form to Eq. (16.17) for the radial wave function in the presence of the Coulomb interaction.

The partial-wave expansion of a plane wave may be written in the form

$$e^{i\underset{\approx}{k}\cdot\underset{\approx}{r}} = \frac{\Gamma(\mu)}{\sqrt{2}}\left(\frac{2}{\kappa}\right)^\lambda \sum_{K=0}^{\infty} i^K (K+\mu) R^{(0)}_{\kappa K}(r) C^\mu_K(\cos\theta_{\underset{\approx}{k}\cdot\underset{\approx}{r}}) \quad . \tag{15.18}$$

Equation (15.10) can be obtained also from (15.18) by taking into account the asymptotic behavior of the radial function $R^{(0)}_{\kappa K}$. As this method of deriving the equation is also of interest for the Coulomb case, we develop it here for a plane wave. Substituting (15.14) into (15.18) and using the fact that

$$C^{\mu}_{K}(x) = (-1)^{K} C^{\mu}_{K}(-x) \quad , \tag{15.19}$$

we obtain

$$e^{i\underset{\sim}{k}\cdot\underset{\sim}{r}} \sim D(\cos\theta_{\underset{\sim}{k}\cdot\underset{\sim}{r}})\left(\frac{2\pi}{i\kappa r}\right)^{\lambda} e^{i\kappa r} + D(-\cos\theta_{\underset{\sim}{k}\cdot\underset{\sim}{r}})\left(\frac{2\pi i}{\kappa r}\right)^{\lambda} e^{-i\kappa r} \quad , \tag{15.20}$$

where

$$D(x) = \frac{\Gamma(\mu)}{2\pi^{\mu+1}} \sum_{K=0}^{\infty} (K+\mu) C^{\mu}_{K}(x) \quad . \tag{15.21}$$

The completeness relation for Gegenbauer polynomials is of the form

$$\sum_{K=0}^{\infty} \frac{1}{h_K} C^{\mu}_{K}(x) C^{\mu}_{K}(x') = \frac{\delta(x-x')}{(1-x^2)^{\mu-1/2}} \quad , \tag{15.22}$$

where h_K is a normalizing factor:

$$h_K = \int_{-1}^{1} [C^{\mu}_{K}(x)]^2 (1-x^2)^{\mu-1/2} dx \quad . \tag{15.23}$$

With the aid of equations from [26], we find

$$\frac{(K+\mu)\Gamma(\mu)}{\pi^{\mu+1}} = \frac{C^{\mu}_{K}(1)\Gamma(\lambda)}{h_K \pi^{\lambda}} \quad . \tag{15.24}$$

Substituting (15.24) into (15.21) and using (15.22) and (7.10), we obtain

$$D(\pm\cos\theta) = \frac{1}{\omega_{n-1}} \cdot \frac{\delta(1\mp\cos\theta)}{(\sin\theta)^{n-3}} \quad . \tag{15.25}$$

If $\cos\theta_1 = \pm 1$, then any function, continuous on the sphere, is independent of the remaining angles, so integration over these angles reduces to multiplication by ω_{n-1}. From (7.9) this implies

$$D(\pm\cos\theta_{\underset{\sim}{k}\cdot\underset{\sim}{r}}) = \delta(\hat{\Omega}_{\underset{\sim}{k}} \mp \hat{\Omega}_{\underset{\sim}{r}}) \quad ; \tag{15.26}$$

thus, (15.20) agrees with (15.10).

§16. *Multidimensional Coulomb Wave Functions*

The solution of the multidimensional Coulomb problem in the form of a product of radial and angular wave functions is given by Kuznetsov [54] and by Bander and Itzykson [55]. The wave functions with the asymptotic behavior "plane wave + scattered wave" are obtained by Peterkop and Shkele [56,57].

We write the Schrödinger equation for the n-dimensional Coulomb problem in the form

$$\left(-\frac{\hbar^2}{2m}\Delta + \frac{\zeta}{r} - E\right)\psi = 0 \quad . \tag{16.1}$$

In terms of the notation

$$Z = -\frac{m\zeta}{\hbar^2} \quad , \quad \kappa^2 = \frac{2mE}{\hbar^2} \quad , \tag{16.2}$$

the Schrödinger equation takes the form

$$\left(\Delta + \frac{2Z}{r} + \kappa^2\right)\psi = 0 \quad . \tag{16.3}$$

We note that $Z > 0$ corresponds to attraction and $Z < 0$ to repulsion.

We look for a solution of Eq. (16.3) in the form of a product

$$\psi = R_{\kappa K}(Z,r)Y_{K\gamma}(\hat{\Omega}) \quad , \qquad (16.4)$$

where $Y_{K\gamma}$ is a hyperspherical harmonic (K harmonic), and γ denotes the set of quantum numbers other than κ and K. When the hyperspherical harmonic depends only on the angle θ_1, defined by (7.1)-(7.4), it takes the form

$$Y_K(\hat{\Omega}) = \sqrt{\frac{(K+\mu)K!(n-3)!\Gamma(\mu)}{2\pi^{n/2}(K+n-3)!}}\; C_K^{\mu}(\cos\theta_1) \quad . \qquad (16.5)$$

The radial Coulomb wave function in the n-dimensional case satisfies the equation

$$\left[\frac{1}{r^{n-1}} \cdot \frac{d}{dr}\left(r^{n-1}\frac{d}{dr}\right) - \frac{K(K+n-2)}{r^2} + \frac{2Z}{r} + \kappa^2\right] R_{\kappa K} = 0 \quad . \qquad (16.6)$$

Its solution is

$$R_{\kappa K} = A_{\kappa K}(2\kappa r)^K e^{-i\kappa r} {}_1F_1(K+\lambda+iW, 2K+2\lambda, 2i\kappa r) \quad , \qquad (16.7)$$

where

$$W = Z/\kappa \quad . \qquad (16.8)$$

Normalizing $R_{\kappa K}$ as $r \to \infty$ by the condition

$$R_{\kappa K} \sim \sqrt{\frac{2}{\pi}}\, r^{-\lambda} \sin\left[\kappa r + W \ln 2\kappa r - (K+\lambda-1)\frac{\pi}{2} + \delta_K\right] \quad , \qquad (16.9)$$

we obtain

$$A_{\kappa K} = \frac{e^{\pi W/2}(2\kappa)^{\lambda}|\Gamma(K+\lambda+iW)|}{\sqrt{2\pi}\,\Gamma(2K+2\lambda)} \quad , \qquad (16.10)$$

$$\delta_K = \arg \Gamma(K+\lambda-iW) \quad . \qquad (16.11)$$

The behavior of the Coulomb wave functions at low energies is of interest. For $\kappa \to 0$ we have (for $Z > 0$)

$$R_{\kappa K} \sim \sqrt{2\kappa r}\, r^{-\lambda} J_{2K+n-2}(\sqrt{8Zr}) \quad , \tag{16.12}$$

where J_n denotes a Bessel function.

Let us now consider the solution of Eq. (16.3) which goes over to a plane wave as $Z \to 0$. By analogy with the three-dimensional solution we shall look for a solution to the n-dimensional problem in the form[1)]

$$\psi_n(Z,\underset{\sim}{k},\underset{\sim}{r}) = C_n e^{i\underset{\sim}{k}\cdot\underset{\sim}{r}} F_n(W,\eta) \quad , \tag{16.13}$$

where C_n is a normalization factor, and η is

$$\eta = \kappa r - \underset{\sim}{k}\cdot\underset{\sim}{r} = 2\kappa r \sin^2\frac{\theta}{2} \quad . \tag{16.14}$$

If $n = 3$, then $\eta = \kappa(r-z)$, i.e., it is proportional to one of the parabolic coordinates. In order to avoid any confusion, we note that here η denotes a different quantity than in Chapter I.

Substituting (16.13) into (16.3), we obtain

$$\left[\eta \frac{d^2}{d\eta^2} + (\lambda - i\eta)\frac{d}{d\eta} + W\right] F_n = 0 \quad . \tag{16.15}$$

A solution of (16.15) is the confluent hypergeometric function

$$F_n = {}_1F_1(iW,\lambda,i\eta) \quad . \tag{16.16}$$

Taking into account the asymptotic behavior of F_n for $\eta \to \infty$, we choose C_n to have the form

$$C_n = e^{\pi W/2} \frac{\Gamma(\lambda - iW)}{\Gamma(\lambda)} \quad . \tag{16.17}$$

---

1) The result is not changed if a solution is sought in the form $\psi_n = C_n e^{i\kappa r} \tilde{F}_n(W,\eta)$.

With this choice of normalization factor the asymptotic behavior of ψ_n is given by the equations

$$\psi_n = \psi_n^{(a)} + \psi_n^{(b)} \quad , \tag{16.18}$$

$$\psi_n^{(a)} = e^{i\underset{\sim}{k}\cdot\underset{\sim}{r} - iW \ln \eta}\, G(iW, iW-\lambda+1, -i\eta) \quad , \tag{16.19}$$

$$\psi_n^{(b)} = \frac{\Gamma(\lambda-iW)}{\Gamma(iW)} (i\eta)^{-\lambda} e^{i\kappa r + iW \ln \eta}\, G(\lambda-iW, 1-iW, i\eta) \quad , \tag{16.20}$$

$$G(\alpha,\beta,z) = 1 + \frac{\alpha\beta}{1!z} + \frac{\alpha(\alpha+1)\beta(\beta+1)}{2!z^2} + \dots \quad . \tag{16.21}$$

Equation (16.19) describes the distorted incident wave, and (16.20) describes the scattered wave. Each of these separately satisfies (16.3). It should be noted that Eqs. (16.19)-(16.21) are applicable for large η; hence, because of (16.14), as $\theta_{\underset{\sim}{k}\cdot\underset{\sim}{r}} \to 0$ the domain of validity in r is removed to infinity.

The asymptotic behavior of the scattered wave agrees with Eqs. (9.1) and (9.2), which were obtained earlier. They differ in an inessential way in that Eq. (16.20) contains a factor η in the argument of the logarithm. From (16.14) it is seen that $\ln \eta$ differs from $\ln r$ by a term which can be included in the phase of the scattering amplitude. If we leave $2\kappa r$ in the argument of the logarithm, the first term of the asymptotic expansion of the scattered wave can be written in the form

$$\psi_n^{(b)} \sim f_n(\theta_{\underset{\sim}{k}\cdot\underset{\sim}{r}}) r^{-\lambda} e^{i\kappa r + iW \ln 2\kappa r} \quad , \tag{16.22}$$

where

$$f_n(\theta) = \frac{\Gamma(\lambda-iW)}{\Gamma(iW)} \left(2\kappa i \sin^2 \frac{\theta}{2}\right)^{-\lambda} \left(\sin^2 \frac{\theta}{2}\right)^{iW} \quad . \tag{16.23}$$

The square of the absolute value of the amplitude can be expressed in terms of elementary functions. For even n=2m we find

$$|f_n(\theta)|^2 = \frac{W\text{th}\pi W}{[2\kappa \sin^2(\theta/2)]^{n-1}} \prod_{t=1}^{m-1} \left[\left(t - \frac{1}{2}\right)^2 + W^2\right] , \qquad (16.24)$$

and for odd $n = 2m+1$

$$|f_n(\theta)|^2 = \frac{W^2}{[2\kappa \sin^2(\theta/2)]^{n-1}} \prod_{t=1}^{m-1} [t^2+W^2] \quad . \qquad (16.25)$$

If $n = 2$ or 3, the product Π is unity. For $n = 3$ (16.25) is simply the Rutherford formula.

For $|W| >> n/2$ we have

$$|f_n(\theta)|^2 \approx \left(\frac{|W|}{2\kappa \sin^2(\theta/2)}\right)^{n-1} , \qquad (16.26)$$

which agrees with the classical expression (26.24). Since

$$W = -\frac{m\zeta}{\hbar^2\kappa} = -\frac{\zeta}{\hbar v} , \qquad (16.27)$$

where v is the velocity, the classical limit is reached for large charges or small velocities. The case $n = 3$ is exceptional, as the quantum and classical differential cross sections coincide for all ζ and v. For other n the quantum and classical differential cross sections are proportional to one another; that is, they differ by a factor independent of θ. Only for $n = 3$ is the quantum differential cross section independent of $\hbar$ for all E.

Similarly, for the square of the absolute value of the normalization factor, we obtain for $n = 2m$

$$|C_n|^2 = \frac{2\pi}{[\Gamma(\lambda)]^2(1+e^{-2\pi W})} \prod_{t=1}^{m-1} \left[\left(t - \frac{1}{2}\right)^2 + W^2\right] , \qquad (16.28)$$

and for $n = 2m+1$

$$|C_n|^2 = \frac{2\pi W}{[\Gamma(\lambda)]^2(1-e^{-2\pi W})} \prod_{t-1}^{m-1} [t^2+W^2] \quad . \qquad (16.29)$$

The behavior of the multidimensional Coulomb wave function as $E \to 0$ is determined for the most part by the behavior of the coefficient C_n. If $Z < 0$ (repulsive case), then as $E \to 0$, C_n falls off exponentially. If $Z > 0$, then as $E \to 0$, C_n goes to infinity while undergoing infinitely rapid oscillation:

$$C_n \sim \frac{\sqrt{2\pi}}{\Gamma(\lambda)} \left(\frac{Z}{i\kappa}\right)^{(n/2)-1} \exp\left[\frac{iZ}{\kappa}\left(1 - \ell n \frac{Z}{\kappa}\right)\right] \quad . \qquad (16.30)$$

For $Z > 0$ and $E \to 0$ the wave function takes the form

$$\psi_n \sim C_n\Gamma(\lambda) \left(\sqrt{2Zr} \sin\frac{\theta}{2}\right)^{1-\lambda} J_{\lambda-1}\left(\sqrt{8Zr} \sin\frac{\theta}{2}\right) \quad . \qquad (16.31)$$

When r is large, but still satisfies $r \ll Z/2E$, we have

$$\psi_n \sim \frac{C_n\Gamma(\lambda) \cos\left[\sqrt{8Zr} \sin\frac{\theta}{2} + \left(1 - \frac{n}{2}\right)\frac{\pi}{2}\right]}{\sqrt{\pi} \left(\sqrt{2Zr} \sin\frac{\theta}{2}\right)^{(n/2)-1}} \quad . \qquad (16.32)$$

Let us now consider the asymptotic form of the incident wave. Equation (16.19) also contains a logarithmic term in the phase, but with a different sign from that in (16.20). In the general case we expect that the Coulomb logarithmic phase shift for a plane wave will differ from the phase shift for a spherical wave. We recall that in the preceding chapters the asymptotic form of the problem in which Z depends on $\hat{\Omega}$ was treated only for a spherical wave. It was assumed that the wave function in the asymptotic domain corresponding to ionization contains only a scattered wave. This followed from the boundary conditions, which stated that at the onset of the collision there were only two free particles -- the incident electron and the atom.

In problems that have three or more unbound, but interacting, particles in the initial state, e.g., in the recombination process, the asymptotic form becomes more complicated because of single and double scattering processes. The asymptotic form of the wave function describing a collision process with three free particles in the initial state has only recently been found by Merkur'ev [58] and Nuttall [59] for short-range forces. The Coulomb case has so far received little attention in the literature.[1)] Just as the logarithmic phase shift of the scattered wave (16.20) is valid in the general case if we consider W to depend on $\hat{\Omega}$, we can expect that the phase shift of the incident wave in the general case of a system of charged particles has a form corresponding to (16.19). The terms describing single and double scattering also apparently differ by a logarithmic phase shift from the corresponding terms in the case of short-range forces. These questions, however, require special consideration (see §11).

Leaving the problem of the asymptotic form of the incident wave when Z depends on $\hat{\Omega}$, we consider in more detail the asymptotic form when Z = const; we have in mind the use of the multidimensional Coulomb wave functions in the limit $r \to \infty$, in integrals of the form

$$\int F(\hat{\Omega}) \psi_n(Z, \underset{\approx}{k}, \underset{\approx}{r}) d\hat{\Omega} \quad . \tag{16.33}$$

The presence of the plane wave in (16.19) suggests the existence of two stationary points: $\theta = 0$ and $\theta = \pi$. Let us

---

1) The Coulomb phase shift of the incident wave is discussed by Rosenberg [176].

consider first the case in which $\theta \approx \pi$. Then, $\eta \approx 2\kappa r$ and Eq. (16.19) is valid for sufficiently large r. Replacing the plane wave by the second term of the asymptotic form (15.10), we obtain

$$\psi_n \sim \delta(\hat{\Omega}_{\underset{\sim}{k}} + \hat{\Omega}_{\underset{\sim}{r}}) \left(\frac{2\pi i}{\kappa r}\right)^{\lambda} e^{-i\kappa r - iW \ln 2\kappa r} \quad . \tag{16.34}$$

In (16.34), as in (16.19), W appears with a negative sign; this is readily understood because (16.34) corresponds to a converging wave, as κr also appears with a negative sign. Earlier we considered the asymptotic form of a spherically diverging wave. The transition to a converging wave is obtained trivially by taking the complex conjugate expression. This expression is also a solution because the Hamiltonian is real. Going to the complex conjugate expression changes the sign of W. Thus, the logarithmic term in (16.19) for $\theta = \pi$ corresponds to the asymptotic behavior of a converging wave, and in this sense it agrees with the results obtained earlier.

If $\theta \approx 0$, then $\eta \approx \kappa r\theta^2/2$. For $\theta \to 0$ the domain of validity of (16.19) and (16.20) is removed to infinity. For large r the quantity η and the wave function both oscillate very rapidly with θ. It remains unclear what kind of asymptotic expression will approximate ψ_n. It is interesting that for $\theta = 0$ the confluent hypergeometric function in (16.13) is equal to unity, and in this case we have a plane wave with no logarithmic phase shift.

To elucidate the asymptotic form with $\theta \approx 0$, we expand the multidimensional Coulomb wave function in partial waves. This is valid if $F(\hat{\Omega})$ in (16.33) contains a finite number of partial harmonics or converges with respect to the harmonics sufficiently rapidly. To obtain this expansion we use the method applied by Gordon [60] in the three-dimensional case.

First we establish the relationship between the Coulomb radial wave function and the free-particle radial wave function:

$$R_{\kappa K} = Be^{-i\delta_K} \int_1^{(0+)} R^{(0)}_{\kappa K}[r(1-t)]e^{i\kappa rt}(-t)^{iW-1}(1-t)^{\lambda-iW-1}dt \quad , \tag{16.35}$$

where

$$B = -e^{\pi W/2}\Gamma(1-iW)/2\pi i \quad . \tag{16.36}$$

The integration contour in (16.35) is a loop beginning and ending at the point $t = 1$ and going around the point $t = 0$ in the positive (counterclockwise) direction. The powers are understood in the sense of the principal value.

It is easy to obtain Eq. (16.35) by using (15.16) for $R^{(0)}_{\kappa K}$ [the complex conjugate expression should be used, but this is trivial since $R^{(0)}_{\kappa K}$ is real; the same is true also of (16.7)], expanding the confluent hypergeometric function in the series

$${}_1F_1(a,c,x) = \frac{\Gamma(c)}{\Gamma(a)} \sum_{n=0}^{\infty} \frac{\Gamma(a+n)x^n}{\Gamma(c+n)n!} \quad , \tag{16.37}$$

integrating term by term according to the formula

$$\int_1^{(0+)} (-t)^{a-1}(1-t)^{b-1}dt = - \frac{2\pi i\Gamma(b)}{\Gamma(1-a)\Gamma(a+b)} \quad , \tag{16.38}$$

summing by means of (16.37), and comparing the result with (16.7) and (16.10).

We next use the representation of the confluent hypergeometric function as a contour integral [26],

$$
{}_1F_1(a,c,x) = -\frac{\Gamma(1-a)\Gamma(c)}{2\pi i\Gamma(c-a)}\int_1^{(0+)} e^{xt}(-t)^{a-1}(1-t)^{c-a-1}dt \quad , \tag{16.39}
$$

which when applied to (16.13) gives

$$
\psi_n = B\int_1^{(0+)} e^{i\kappa r(1-t)\cos\theta} e^{i\kappa rt}(-t)^{iW-1}(1-t)^{\lambda-iW-1}dt \quad . \tag{16.40}
$$

We expand the first exponential in (16.40) in the series (15.18), where r is replaced by r(1-t); then we use (16.35). As a result we obtain[1)]

$$
\psi_n = \frac{\Gamma(\mu)}{\sqrt{2}}\left(\frac{2}{\kappa}\right)^{\lambda}\sum_{K=0}^{\infty} i^K(K+\mu)e^{i\delta_K}R_{\kappa K}(r)C_K^{\mu}(\cos\theta_{\underset{\sim}{k}\cdot\underset{\sim}{r}}) \quad . \tag{16.41}
$$

Since $C_k^{1/2} = P_K$ (P_K is a Legendre polynomial), for n = 3 this expression agrees with the well-known expansion of the three-dimensional Coulomb wave function in terms of Legendre polynomials.

In order to determine the asymptotic behavior of ψ_n, we replace the radial wave function in (16.41) by its asymptotic expression (16.9). As the number K is increased, the domain of validity of (16.9) recedes, so the use of (16.9) is justified if the integral (16.33) contains only a finite number of harmonics. This applies also to (15.14), (15.20), and (15.10). In a manner analogous to the derivation of Eqs. (15.20) and (15.26) we obtain

---

1) Equation (16.41) also follows from the addition formula for the confluent hypergeometric function given in [26].

$$\psi_n \sim \tilde{f}_n(\theta_{\underset{\approx}{k}\cdot\underset{\approx}{r}}) r^{-\lambda} e^{i\kappa r + iW \ln 2\kappa r} + \delta(\hat{\Omega}_{\underset{\approx}{k}} + \hat{\Omega}_{\underset{\approx}{r}}) \left(\frac{2\pi i}{\kappa r}\right)^{\lambda} e^{-i\kappa r - iW \ln 2\kappa r} , \tag{16.42}$$

where

$$\tilde{f}_n(\theta) = \sum_{K=0}^{\infty} \tilde{a}_K C_K^{\mu}(\cos\theta) , \tag{16.43}$$

$$\tilde{a}_K = \frac{\Gamma(\mu)}{2\sqrt{\pi}} \left(\frac{2}{i\kappa}\right)^{\lambda} (K+\mu) e^{2i\delta_K} . \tag{16.44}$$

The second term in (16.42) is identical to (16.34). As a function of r, the first term corresponds to the general asymptotic form of a spherically diverging wave. The function $\tilde{f}_n(\theta)$, which gives the dependence on θ, is in a certain sense an expansion of the scattering amplitude $f_n(\theta)$, defined by (16.23), in Gegenbauer polynomials. The expansion coefficients are defined by the integral

$$a_K = \frac{1}{h_K} \int_0^{\pi} f_n(\theta) C_K^{\mu}(\cos\theta)(\sin\theta)^{n-2} d\theta , \tag{16.45}$$

where h_K is found from (15.23). Substituting (16.23) into (16.45) we obtain

$$a_K = \frac{\Gamma(\lambda - iW)}{2h_K \Gamma(iW)} \left(\frac{2}{i\kappa}\right)^{\lambda} \int_0^{\pi} C_K^{\mu}(\cos\theta) \left(\cos\frac{\theta}{2}\right)^{n-2} \times$$

$$\times \left(\sin\frac{\theta}{2}\right)^{2iW-1} d\theta . \tag{16.46}$$

The integral in (16.46) does not converge owing to the singularity at $\theta = 0$. If we replace the lower limit in the integral by $\Delta > 0$, then as $\Delta \to 0$, the integral oscillates and does not approach any definite limit. However, the following regularization of the integral is possible. To the exponent 2iW-1 in (16.46) we add $\varepsilon > 0$ and after calculating the integral we take the limit $\varepsilon \to 0$. In this calculation it is convenient to express the Gegenbauer polynomial in terms of the hypergeometric function. From [26] we have

$$\frac{1}{h_K} C_K^{\mu}(\cos\theta) = \frac{(K+\mu)\Gamma(\mu)}{\sqrt{\pi}\,\Gamma(\lambda)}\, {}_2F_1\left(-K, K+2\mu, \lambda, \sin^2\frac{\theta}{2}\right) \quad . \qquad (16.47)$$

Then we transform to the integration variable $x = \sin^2\theta/2$, expand the hypergeometric function in a power series in x (a finite series), integrate term by term, and then sum according to Zaalschutz' formula [26, §2.1.5]. We find as a result

$$a_K = \tilde{a}_K \quad . \qquad (16.48)$$

Equations (16.43) and (16.44) differ from the usual partial wave expansions of the amplitude in that (16.44) contains the term $\exp(2i\delta_K)$ instead of the difference $\exp(2i\delta_K)-1$. Accordingly, the factor $\delta(\hat{\Omega}_{\underset{\sim}{k}} - \hat{\Omega}_{\underset{\sim}{r}})$ is not separated out in the first term of (16.42). When $W \to 0$,

$$f_n \to 0 \quad ; \quad \tilde{f}_n \to \left(\frac{2\pi}{i\kappa}\right)^{\lambda} \delta(\hat{\Omega}_{\underset{\sim}{k}} - \hat{\Omega}_{\underset{\sim}{r}}) \quad . \qquad (16.49)$$

For $K\to\infty$ the Coulomb phase shift diverges logarithmically ($\delta_K \sim -W \ln K$), so the series (16.43) does not converge, nor does the series in which $\exp(2i\delta_K)$ is replaced by $\exp(2i\delta_K) - 1$ [61]. In practical calculations the coefficients of the divergent series for $f_n(\theta)$ can be used to determine the coefficients of a convergent series for the product $(1-\cos\theta)^m f_n(\theta)$, where $m = 1,2,\ldots$ [62], or they can be used in other summation methods. As already noted, the asymptotic form (16.42) is applicable if, for all practical purposes, the integral (16.33) contains a finite number of harmonics. There is then no problem with the convergence of the amplitude $\tilde{f}_n$. The numerical value of $\tilde{f}_n$ is not important, since only the second term of (16.42) contributes to the integrals usually considered.

The asymptotic form (16.18)-(16.20) is of the type "incident wave + diverging wave." In constructing integral representations of the ionization amplitude, we use the function

$$\psi_n^{(-)}(Z,\underset{\sim}{k},\underset{\sim}{r}) = \psi_n^*(Z,-\underset{\sim}{k},\underset{\sim}{r}) \quad , \qquad (16.50)$$

for which Eqs. (16.18)-(16.20) take the form "incident wave + converging wave." Equations (16.42) and (16.50) imply

$$\psi_n^{(-)} \sim \delta(\hat{\Omega}_{\underset{\sim}{k}} - \hat{\Omega}_{\underset{\sim}{r}}) \left(\frac{2\pi}{i\kappa r}\right)^{\lambda} e^{i\kappa r + iW \ell n\, 2\kappa r} +$$
$$+ \tilde{f}_n^*(-\cos\theta_{\underset{\sim}{k}\cdot\underset{\sim}{r}}) r^{-\lambda} e^{-i\kappa r - iW \ell n\, 2\kappa r} \quad . \qquad (16.51)$$

The function $\psi_n^{(-)}$ is convenient in that it has a simpler factor multiplying the diverging wave.

In integral representations of the ionization amplitude, products of Coulomb wave functions depending on the different coordinates in the multidimensional space are also used. The asymptotic behavior of a product of Coulomb wave functions is also of the form (16.51). We demonstrate this in the example of the product of two three-dimensional Coulomb functions:

$$\Phi = \psi_3^{(-)}(Z_1,\underset{\sim}{k}_1,\underset{\sim}{r}_1)\psi_3^{(-)}(Z_2,\underset{\sim}{k}_2,\underset{\sim}{r}_2) \quad , \quad k_1^{\;2} + k_2^{\;2} = \kappa^2 \quad . \qquad (16.52)$$

A product of two three-dimensional expressions of the form (16.51) contains four terms. We use the two-dimensional version of Eq. (15.10) which gives

$$e^{\pm ik_1r_1 \pm ik_2r_2} \sim \delta(\alpha_k - \alpha_r)\sqrt{\frac{2\pi}{\pm i\kappa r}}\, e^{\pm i\kappa r} \quad , \tag{16.53}$$

where

$$\alpha_k = \operatorname{arctg}\frac{k_2}{k_1} \quad ; \quad \alpha_r = \operatorname{arctg}\frac{r_2}{r_1} \quad . \tag{16.54}$$

In (16.53) we have considered only the first term of (15.10). The second stationary point, for which $\alpha_r = -\alpha_k$, does not occur, because r_1 and r_2, as well as k_1 and k_2, take positive values. For this reason terms containing $\exp(\pm ik_1r_1 \mp ik_2r_2)$ have no stationary points at all. As a result we obtain

$$\Phi \sim \delta(\hat{\Omega}_{\underset{\sim}{k}} - \hat{\Omega}_{\underset{\sim}{r}})\left(\frac{2\pi}{i\kappa r}\right)^{5/2} e^{i\kappa r + iw\,\ln\,\kappa r + i\beta} +$$

$$+ F(\hat{\Omega}_{\underset{\sim}{k}}, \hat{\Omega}_{\underset{\sim}{r}})\, r^{-5/2} e^{-i\kappa r - iw\,\ln\,\kappa r - i\beta} \quad , \tag{16.55}$$

where

$$w = \frac{Z_1}{k_1} + \frac{Z_2}{k_2} \quad , \tag{16.56}$$

$$\beta = \frac{Z_1}{k_1}\ln\frac{2k_1^{\,2}}{\kappa^2} + \frac{Z_2}{k_2}\ln\frac{2k_2^{\,2}}{\kappa^2} \quad , \tag{16.57}$$

$$F(\hat{\Omega}_{\underset{\sim}{k}}, \hat{\Omega}_{\underset{\sim}{r}}) = \sqrt{\frac{2\pi i}{\kappa}}\, \tilde{f}_3^*(-\cos\theta_{\underset{\sim}{k}_1\underset{\sim}{r}_1})\tilde{f}_3^*(-\cos\theta_{\underset{\sim}{k}_2\underset{\sim}{r}_2})\, \frac{\delta(\alpha_k - \alpha_r)}{\sin\alpha_k\,\cos\alpha_k} \quad . \tag{16.58}$$

In the asymptotic form (16.55) we have used the relation

$$\delta(\hat{\Omega}_{\underset{\sim}{k}} - \hat{\Omega}_{\underset{\sim}{r}}) = \delta(\hat{\Omega}_{\underset{\sim}{k}1} - \hat{\Omega}_{\underset{\sim}{r}1})\,\delta(\hat{\Omega}_{\underset{\sim}{k}2} - \hat{\Omega}_{\underset{\sim}{r}2})\, \frac{\delta(\alpha_k - \alpha_r)}{\sin^2\alpha_k\,\cos^2\alpha_k} \quad , \tag{16.59}$$

which follows from (3.16).

CHAPTER V

INTEGRAL REPRESENTATIONS OF THE IONIZATION AMPLITUDE

§17. *Derivation of the Integral Representation*

In §5 it was shown that for short-range forces the ionization amplitude can be represented as an integral (5.7), the integrand of which contains certain well-known functions and the exact solution of the Schrödinger equation, which is in general unknown. Knowing the asymptotic form of the wave function, we can construct similar expressions for the case of three particles with Coulomb interactions [6,19]. On the basis of these integral expressions we can reach some conclusions about certain properties of the amplitude, e.g. its behavior for $E \to 0$. The integral expressions find application in the determination of the ionization amplitude by extrapolation from the complex energy domain [63]. The construction of a variational principle for the ionization problem [52,64] is also related to these expressions.

Let us consider the integral

$$I = \int \Psi(\underset{\sim}{r}_1, \underset{\sim}{r}_2)(H - E)\Phi^*(\underset{\sim}{r}_1, \underset{\sim}{r}_2) d\underset{\sim}{r}_1 d\underset{\sim}{r}_2 \quad , \qquad (17.1)$$

where Ψ is the exact solution to the Schrödinger equation which satisfies conditions (1.3) and (1.4), and Φ is some nonsingular function satisfying certain asymptotic conditions, which will be specified later.

The integrand of (17.1) can be expressed in a form involving the six-dimensional divergence operator. Using

$$(H - E)\Psi = 0 \quad , \quad H = -\frac{1}{2}\Delta - \frac{Z(\hat{\Omega})}{r} \quad , \tag{17.2}$$

where Δ denotes the Laplacian in six-dimensional space, we can write

$$\Psi(H - E)\Phi^* = \Psi(H - E)\Phi^* - \Phi^*(H - E)\Psi \tag{17.3}$$

$$= \frac{1}{2}(\Phi^*\Delta\Psi - \Psi\Delta\Phi^*) = \frac{1}{2}\nabla\cdot(\Phi^*\nabla\Psi - \Psi\nabla\Phi^*) \quad , \tag{17.4}$$

where ∇ denotes the six-dimensional gradient operator. The integral involving the divergence operator can be expressed in terms of an integral over a hypersphere:

$$I = \frac{1}{2}\lim_{r\to\infty} r^5 \int \left(\Phi^* \frac{\partial\Psi}{\partial r} - \Psi \frac{\partial\Phi^*}{\partial r}\right) d\hat{\Omega} \quad . \tag{17.5}$$

In Eq. (17.5) we replace the function Ψ by the asymptotic form (6.39). It will be shown later that the discrete part of (1.3) and (1.4) does not contribute to the integral. In order that the ionization amplitude for a definite direction $\hat{\Omega}_{\underset{\approx}{k}}$ be separated from the integral over $\hat{\Omega}$, the function Φ should have a singularity of the type $\delta(\hat{\Omega} - \hat{\Omega}_{\underset{\approx}{k}})$ as $r \to \infty$; in order that (17.5) converge to a finite limit, Φ as a function of r should have the same asymptotic form as Ψ. Thus, we arrive at the asymptotic condition

$$\Phi \sim \text{const} \cdot \delta(\hat{\Omega}_{\underset{\approx}{r}} - \hat{\Omega}_{\underset{\approx}{k}}) r^{-5/2} e^{i\kappa r + [iZ(\hat{\Omega}_{\underset{\approx}{k}})/\kappa] \ln \kappa r} \quad . \tag{17.6}$$

The function containing only the diverging wave in the asymptotic domain usually turns out to be singular at $r = 0$. However, it is easy to see that if a converging wave is added to (17.6), then (17.5) is not changed, because the additional terms in (17.5) cancel one another. Henceforth we shall assume that Φ has the asymptotic form (16.55), where the quantities w, β, and F are not necessarily given by Eqs. (16.56)-(16.58), although w must be real. Of this form are the asymptotic expressions for a six-dimensional plane wave, a six-dimensional Coulomb wave function, and the product of Coulomb wave functions of lower dimension expressed in terms of the coordinates of the six-dimensional space. In order that (17.5) converge to a finite limit, it is necessary that

$$w = \frac{Z(\hat{\Omega}_{\underset{\sim}{k}})}{\kappa} \equiv W(\hat{\Omega}_{\underset{\sim}{k}}) \quad . \tag{17.7}$$

With this condition we find

$$A_{00}(\hat{\Omega}_{\underset{\sim}{k}}) = \kappa^{3/2}(2\pi)^{-5/2}e^{i\beta+i\pi/4}\int \Psi(H-E)\Phi^* d\underset{\sim}{r}_1 d\underset{\sim}{r}_2 \quad . \tag{17.8}$$

Note that an asymptotic condition of the form (16.55) means that Φ must have the asymptotic form "incident wave + converging wave" in six-dimensional space. In other words, Φ must have the asymptotic form of an incident wave corresponding to the case in which one of the ionization channels is the entrance channel.

If Φ satisfies the equation

$$(H - V_\Delta - E)\Phi = 0 \quad , \tag{17.9}$$

then (17.8) can be rewritten in the form

$$A_{00}(\hat{\Omega}_{\underset{\sim}{k}}) = \kappa^{3/2}(2\pi)^{-5/2}e^{i\beta+i\pi/4}\int \Psi V_\Delta \Phi^* d\underset{\sim}{r}_1 d\underset{\sim}{r}_2 \quad . \tag{17.10}$$

For short-range forces, when $\beta = 0$, Eq. (17.10) agrees with the integral expression obtained in §5.

The asymptotic form (6.39) is applicable if both r_1 and $r_2 \to \infty$. If one of them remains finite, then the behavior of the wave function is determined by Eq. (1.3) and (1.4), from which it is seen that the terms corresponding to excitation of the levels of the discrete spectrum fall off more slowly than $r^{-5/2}$. Therefore, their influence in the integral (17.1) needs to be ascertained. For this we note that the six-dimensional expression $\nabla\cdot(\Psi\nabla\Phi)$ is a sum of two three-dimensional expressions corresponding to the coordinates $\underset{\sim}{r}_1$ and $\underset{\sim}{r}_2$. From (17.4) we have

$$\Psi(H-E)\Phi^* = \frac{1}{2}\sum_{j=1}^{2} \nabla_j\cdot(\Phi^*\nabla_j\Psi - \Psi\nabla_j\Phi^*) \quad , \qquad (17.11)$$

where ∇_j is the three-dimensional operator acting on $\underset{\sim}{r}_j$. Transforming (17.1) to integrals over surfaces in three-dimensional space and retaining only the discrete terms, we obtain

$$I_d = \frac{1}{2}\sum_{j=1}^{2} \lim_{r_j\to\infty} r_j^{\,2} \int\!\!\int \left[\Phi^* \frac{\partial\Psi}{\partial r_j} - \Psi\frac{\partial\Phi^*}{\partial r_j}\right] d\hat{\Omega}_j d\underset{\sim}{r}_t \quad , \qquad (17.12)$$

where $t = 2$ if $j = 1$ and $t = 1$ if $j = 2$. The notation I_d indicates that in (17.12) only the sums in (1.3) and (1.4) over the discrete spectrum of the atom are included in the asymptotic behavior of Ψ; these sums are not considered when (6.39) is used for the asymptotic behavior of Ψ in (17.5).

In connection with the slow decay of the asymptotic forms (1.3) and (1.4) we must determine more precisely the asymptotic behavior of Φ when one of r_1 and r_2 remains finite. We shall assume that for $r_j\to\infty$

$$\Phi \sim \chi_t(\underset{\sim}{r}_t) r_j^{-1}\left[f_{j1}(\hat{\Omega}_j) e^{i\kappa_j r_j + iw_j \ln\kappa_j r_j} + f_{j2}(\hat{\Omega}_j) e^{-i\kappa_j r_j - iw_j \ln\kappa_j r_j}\right], \qquad (17.13)$$

where

$$\kappa_1^{\,2} + \kappa_2^{\,2} = 2E \quad , \quad 0 \le \kappa_j \le \sqrt{2E} \quad . \qquad (17.14)$$

In (17.13) we assume that χ_t is nonsingular; f_{j1} and f_{j2} can have δ-function singularities. The condition that $\kappa_j \le \sqrt{2E}$ means that Φ has no components corresponding to a transition of the atom to the discrete states. This puts definite restrictions on V_Δ. In particular, it excludes $V_\Delta = 0$. This restriction implies that we are considering only simple forms of Φ. A broader class of functions Φ is used in the formulation of the variational principle for the ionization amplitude in the case of short-range forces [52,64] (see §20).

If we substitute (1.3), (1.4), and (17.13) into (17.12), it turns out that none of the terms of the sum over discrete atomic states goes to a definite limit for $r_j \to \infty$, but rather that they all oscillate, because k_n is not equal to either κ_1 or κ_2. According to (17.14) $\kappa_j \le \sqrt{2E}$, but from (1.8) it is seen that $k_n > \sqrt{2E}$.

The lack of convergence in Eq. (17.12) is not a consequence of the Coulomb interaction. It occurs also with short-range forces. This difficulty can be avoided by requiring that χ_t in (17.13) be orthogonal to all atomic wave functions of the discrete spectrum. Then

$$I_d = 0 \quad . \tag{17.15}$$

Physically it seems preferable to apply regularization methods, as, for example, the introduction of wave packets. The replacement of (17.13) by a wave packet in κ_j also leads to (17.15). A similar uncertainty occurs in the exchange part of the ionization amplitude in the Born-Oppenheimer approximation, if we use the post form; this is usually removed by going to the prior form, which is justified by introducing wave packets [65]. However, the wave-packet method also implies (17.15) when $\kappa_j = k_n$. Lieber, Rosenberg, and Spruch [52] have applied the Dirac regularization method, in which an average is taken over a large integral in r_j, to an expression of the form (17.12); this also gives (17.15) when $\kappa_j \neq k_n$. We note that due to the randomness of the phases, the interference of the terms that correspond to different atomic states implies that (17.12) is by itself close to zero. Henceforth we shall assume $I_d = 0$.

We now examine some choices for the function Φ. Essentially we may say that Φ describes the final state of the ionization process. In the Born approximation the appropriate function is

$$\Phi = e^{i\underset{\sim}{k}_1 \cdot \underset{\sim}{r}_1} \psi_3^{(-)}(1,\underset{\sim}{k}_2,\underset{\sim}{r}_2) \quad , \qquad (17.16)$$

where $\psi_3^{(-)}$ is the three-dimensional Coulomb wave function that describes the motion of an electron in the field of a fixed proton. However, in the general case the function (17.16) does not satisfy (17.7); consequently, as $r \to \infty$, Eq. (17.5) oscillates logarithmically (in the Born approximation the integral converges, because Ψ is replaced by an approximate expression). Regularization methods cannot be applied in this case. The divergence is of a specifically Coulomb nature, and the natural interference of the various terms is absent. Therefore, Eq. (17.16) must be regarded as unacceptable. We note, however, that (17.5) converges in modulus if (17.16) is used; the phase of the integral remains indeterminate.

In order to obtain an expression that converges also in the phase, we can use a product of Coulomb functions that describe the motion of an electron in the fields produced by charges that are not necessarily unity:

$$\Phi = \psi_3^{(-)}(Z_1,\underset{\sim}{k}_1,\underset{\sim}{r}_1)\psi_3^{(-)}(Z_2,\underset{\sim}{k}_2,\underset{\sim}{r}_2) \, , \quad k_1^2 + k_2^2 = \kappa^2 \quad . \qquad (17.17)$$

Equation (17.7) together with (16.56) and (9.9) gives the condition on the charges Z_1 and Z_2

$$\frac{Z_1}{k_1} + \frac{Z_2}{k_2} = \frac{1}{k_1} + \frac{1}{k_2} - \frac{1}{|\underset{\sim}{k}_1 - \underset{\sim}{k}_2|} \quad . \tag{17.18}$$

One of the charges can be chosen arbitrarily. Comparing (17.9) with (16.3), we find that corresponding to the function (17.17), the appropriate potential is

$$V_\Delta = \frac{Z_1 - 1}{r_1} + \frac{Z_2 - 1}{r_2} + \frac{1}{r_{12}} \quad . \tag{17.19}$$

Note that according to (17.18) Z_1 and Z_2 depend on the direction for which the ionization amplitude is being considered. This direction is determined by the vectors $\underset{\sim}{k}_1$ and $\underset{\sim}{k}_2$. In this direction in the space of $\underset{\sim}{r}_1$ and $\underset{\sim}{r}_2$ (i.e., for $\underset{\sim}{r}_i = \underset{\sim}{k}_i r/\kappa$) the right side of Eq. (17.19) goes to zero, which ensures the convergence of the integral in (17.10).

Another possible choice for Φ is the six-dimensional Coulomb wave function determined by the equation

$$\left(\Delta_1 + \Delta_2 + \frac{2Z}{r} + \kappa^2\right)\Psi_6^{(-)}(Z;\underset{\sim}{k}_1,\underset{\sim}{k}_2;\underset{\sim}{r}_1,\underset{\sim}{r}_2) = 0 \quad . \tag{17.20}$$

The condition (17.7) in this case takes the form

$$\frac{Z}{\kappa} = \frac{1}{k_1} + \frac{1}{k_2} - \frac{1}{|\underset{\sim}{k}_1 - \underset{\sim}{k}_2|} \quad . \tag{17.21}$$

§18. *The Limit E → 0 in the Integral Representation*

The integral representations can be used to determine the dependence of the amplitude on the energy E as E → 0. For this purpose we take the limit E → 0 under the integral sign;

however, it must be borne in mind that this limit is justified only under certain conditions.

Besides (17.17) and (17.20) other combinations of Coulomb wave functions can also be used for Φ, for example, the product of three two-dimensional Coulomb functions corresponding to the charges involved, or the product of a four-dimensional and a two-dimensional Coulomb function. These are of the asymptotic type (16.55) and can be made to satisfy (17.7) by the choice of the charges. By themselves such functions have no immediate physical meaning, but they serve to show that taking the limit $E \to 0$ under the integral sign in (17.1) gives to the ionization amplitude a different energy dependence when different functions Φ are used. Thus, interchanging the limit $E \to 0$ and the integration in (17.1) is not always permissible.

In this limiting procedure it is assumed that Ψ, the solution of the Schrödinger equation, goes to a finite limit as $E \to 0$. This is justified by the fact that the main parameter determining the solution -- the incident-electron momentum -- remains finite in the limit $E \to 0$. When the electron collides with the hydrogen atom in the ground state, $k_o \to 1$ as $E \to 0$. That the solution depends regularly on this parameter has been proved for ordinary differential equations (Poincaré's theorem) [66]. An equivalent assumption is used in the Wigner threshold theory [67]. If $\Psi \to$ const, then the energy dependence of the integrand in (17.1) or (17.10) is determined by the function Φ with the parameters $k_1, k_2 \to 0$.

The energy dependence of the Coulomb wave functions is given by Eqs. (16.30)-(16.32). For Φ given by (17.16) we obtain in the limit $E \to 0$

$$\Phi \sim k_2^{-1/2} \sim E^{-1/4} \quad . \qquad (18.1)$$

Taking the limit E → 0 under the integral sign in (17.10) in this case we are led to

$$A_{00} \sim E^{1/2} \quad . \tag{18.2}$$

Substituting this result into Eq. (3.15) for the effective cross section we obtain

$$\sigma \sim E^{3/2} \quad , \tag{18.3}$$

which agrees with the Born-approximation result. Since the cross section is expressed in terms of the absolute value of the amplitude, the oscillations in (16.30) are not important.

If a product of three-dimensional Coulomb wave functions (17.17) is used for Φ, then

$$\Phi \sim k_1^{-1/2} k_2^{-1/2} \sim E^{-1/2} \quad , \tag{18.4}$$

which leads to a linear threshold law

$$\sigma \sim E \quad . \tag{18.5}$$

If we choose for Φ the six-dimensional Coulomb function, then we find an unexpected result:

$$\Phi \sim E^{-1} \; , \; \sigma \sim \text{const} \quad . \tag{18.6}$$

In contrast with this the product of two-dimensional functions gives

$$\Phi \sim \text{const} \; , \; \sigma \sim E^2 \quad . \tag{18.7}$$

The linear ionization threshold law (18.5) was first obtained by Geltman [68] in an approximate calculation, which differed from the Born approximation in that both electrons were described in the final state by Coulomb wave functions with $Z_1 = Z_2 = 1$. Since the function Ψ was replaced by an approximate expression, the integral (17.1) turned out to be convergent. The theoretical justification of the linear threshold law given by Rudge and Seaton [39] involved taking the limit under the integral sign in (17.1), using the exact wave function for Ψ and Eq. (17.17) for Φ with the condition (17.18). The difference in the results (18.5)-(18.7) indicates that convergence of the integral is insufficient to justify interchanging the limit and the integration.

The inadmissibility of interchanging these two limiting processes results from the growth of the effective domain of integration in (17.1) as $E \to 0$. The boundary of the configuration-space domain contributing to (17.1) is determined by the condition

$$(H - E)\Phi \approx 0 \quad . \tag{18.8}$$

For the various functions Φ considered here the limit of validity of the condition (18.8) coincides with that of the asymptotic form (16.55), and both are removed to infinity as $E \to 0$. On the other hand, the dependence of Φ on E is not uniform for all r, and the dependence on κ of the asymptotic form (16.55) is different from that in (16.30).

We may expect that the interchange of the limit and integration in (17.1) will be justified if the function Φ is chosen so that the boundary of the domain of validity of the condition (18.8) goes to a finite limit as $E \to 0$. Then, for all practical purposes, we are dealing with an integral over a finite volume. The wave function in the semiclassical approximation has the required property. In §27 it will be shown that this approximation leads to the threshold law $\sigma \sim E^{1.127}$, which was obtained by Wannier [50] using classical methods.

If V_Δ in (17.10) is a short-range potential, then for all practical purposes the domain of integration has a finite volume. Taking the limit under the integral sign should then again lead to the correct threshold dependence.

Let us consider several examples. If all the particles interact with short-range potentials, for Φ we can take the product of plane waves

$$\Phi = \exp(i\mathbf{k}_1\cdot\mathbf{r}_1)\exp(i\mathbf{k}_2\cdot\mathbf{r}_2) \quad . \tag{18.9}$$

Then $\Phi \to 1$ as $E \to 0$, and the limit gives

$$\sigma \sim E^2 \quad , \tag{18.10}$$

which is the correct threshold dependence for the disintegration cross section of a deuteron in collision with a neutron [48,69].

The Born-approximation result (18.3) is correct for the ionization of an atom by a neutral particle. When Φ is given by (17.16), V_Δ is a short range potential.

The linear threshold dependence (18.5) would be correct in the case of a short-range interaction between electrons and a Coulomb interaction between the electrons and the nucleus. Then, Φ in the form (17.17) with $Z_1 = Z_2 = 1$ would be applicable.

The opposite situation is observed in the ionization of a negative ion by an electron. Then, in the final state there is only a Coulomb interaction between the outgoing electrons. We obtain a short-range potential V_Δ by choosing Φ in the form of a product, similar to (11.3),

$$\Phi = \exp\left[\frac{i}{2}(\mathbf{k}_1+\mathbf{k}_2)\cdot(\mathbf{r}_1+\mathbf{r}_2)\right]\psi_3^{(-)}\left(-\frac{1}{2}, \frac{\mathbf{k}_1-\mathbf{k}_2}{2}, \mathbf{r}_1-\mathbf{r}_2\right), \tag{18.11}$$

which describes the free motion of the center of mass of the electrons and the relative Coulomb motion. In contrast with the other cases considered, we now find a Coulomb repulsion. Taking into account the exponential decay of the normalization constant given by Eq. (16.29), we obtain for this case the result of Rudge [70]

$$\sigma \sim E^{3/2} \exp\left(-\frac{\text{const}}{\kappa}\right) \quad . \tag{18.12}$$

The examples discussed here correspond to the cases in which the Schrödinger equation admits a separation of variables in the asymptotic domain where short-range potentials can be ignored, so that analytical solutions can be found.

§19. *Method of Complex Energies*

On the basis of the integral representation of the ionization amplitude, McCartor and Nuttall [63] have developed a method of calculating the ionization amplitude by extrapolation from the complex-energy domain. In essence, this method involves the use of the integral representation (17.8) and (17.10) to determine the amplitude for Im E > 0. The vectors $\underset{\sim}{k}_1$ and $\underset{\sim}{k}_2$, determining Φ, remain real (i.e., Φ does not change), and Ψ is a solution of the Schrödinger equation with complex energy E. In Eqs. (17.1) and (17.8) E should be replaced by $\kappa^2/2 = (k_1^2 + k_2^2)/2$. The numerical calculation of Ψ for Im E > 0 turns out to be less laborious than for a real energy. The amplitude is calculated for several complex energies and the results are extrapolated to the real axis.

McCartor and Nuttall [63] considered two choices for Φ. One of them is identical with (17.17) and (17.18), and the other is the product of plane waves (18.9). The wave function (18.9) is convenient in that it has a simple form. Using this function is equivalent to defining the T matrix in the momentum representation. The integral (17.10) with this function does not converge for real energies, but does converge for Im E > 0. McCartor and Nuttall showed that the integral has a logarithmic singularity as Im E → 0 (a similar result was obtained by Veselova [34] using a different method):

$$I(\hat{\Omega}) \sim F(\hat{\Omega})(k-\kappa)^{-iW(\hat{\Omega})} A_{00}(\hat{\Omega}) \quad , \qquad (19.1)$$

where $k = \sqrt{2E}$, F is a nonsingular function, A_{00} denotes the amplitude for Im E = 0, and the direction $\hat{\Omega}$ corresponds to the vectors $\underset{\sim}{k}_1$ and $\underset{\sim}{k}_2$. All the quantities except A_{00} and I are known; so that if we calculate I, we can find A_{00}.

By using the expansion (6.5) McCartor and Nuttall concluded that the higher terms in the expansion of the integral (17.1) are of a form analogous to (6.5):

$$I(\hat{\Omega}) \sim (k-\kappa)^{-iW} \sum_{n=0}^{\infty} \sum_{m=0}^{2n} b_{nm}(\hat{\Omega})(k-\kappa)^n [\ell n(k-\kappa)]^m \quad . \qquad (19.2)$$

They showed that $b_{12} = 0$.

In the version using (17.17) and (17.18) the integral I is not singular for Im E = 0, but the higher expansion terms in this case also contain powers of $\ell n(k-\kappa)$. Thus, the behavior of the amplitude as a function of E in the neighborhood of the real axis in many ways is analogous to the behavior of the wave function in coordinate space for $r \to \infty$.

The applicability of the extrapolation method, when a removable singularity is present, was verified by McCartor and Nuttall for the two-particle Coulomb scattering amplitude.

§20. *The Variational Principle in the Ionization Problem*

The Kohn variational principle [71] for break-up processes similar to ionization has been formulated for short-range forces by a number of authors [52,64,174,175]. In the Coulomb case the asymptotic behavior of the incident wave in the process of recombination has not been given sufficient attention.[1) In the present section we shall consider the solution of Eq. (1.1), assuming that (1.2) contains only short-range potentials; for example, r^{-1} is replaced by $e^{-cr}r^{-1}$, where $c > 0$.

In constructing a variational principle in addition to a wave function that directly describes the process of ionization, i.e., that satisfies the boundary conditions (1.3) and (1.4), it is necessary also to consider the solution of Eq. (1.1) that describes the inverse process of recombination. To investigate the asymptotic form of this solution of the Schrödinger equation, the Faddeev equations [72] are most convenient. By means of these equations it can be shown [58,59] that the wave function for the recombination process can be represented in the form

$$\Phi = \Phi_0 + \Phi_1 + \Phi_2 + \Phi_3 \quad . \tag{20.1}$$

Here Φ_0 describes the initial state

$$\Phi_0 = e^{i\underset{\sim}{k}_1\cdot\underset{\sim}{r}_1 + i\underset{\sim}{k}_2\cdot\underset{\sim}{r}_2} \; , \; k_1{}^2 + k_2{}^2 = \kappa^2 \quad , \tag{20.2}$$

Φ_1 corresponds to a single scattering of the electrons by the nucleus or among themselves; i.e., it contains terms with the asymptotic forms

---

1) The Coulomb case is treated by Rosenberg [176].

$$e^{i\underset{\sim}{k}_1\cdot\underset{\sim}{r}_1} r_2^{-1} t_2(\hat{\Omega}) e^{ik_2 r_2} ,$$

$$e^{(i/2)(\underset{\sim}{k}_1+\underset{\sim}{k}_2)\cdot(\underset{\sim}{r}_1+\underset{\sim}{r}_2)} r_{12}^{-1} t_{12}(\hat{\Omega}_{12}) e^{ik_{12}r_{12}} , \qquad (20.3)$$

which fall off as r^{-1}, and Φ_2 corresponds to the double processes, for example, the process in which one of the electrons is first scattered by the nucleus and then by the other electron. The terms of Φ_2 fall off as r^{-2} or faster. The coefficients of the asymptotic form of Φ_2 are determined by the amplitudes for the two-particle processes.

Φ_3 describes the "true" three-particle processes. Its asymptotic form contains three-particle scattering terms, which fall off as $r^{-5/2}$,

$$\Phi_3 \underset{\substack{r_1\to\infty \\ r_2\to\infty}}{\widetilde{\quad}} \tilde{A}(\hat{\Omega}) r^{-5/2} e^{i\kappa r} \qquad (20.4)$$

and, in addition, terms that correspond to recombination,

$$\Phi_3 \underset{r_1\to\infty}{\widetilde{\quad}} r_1^{-1} \sum_{n\ell m} \tilde{f}_{n\ell m}(\hat{\Omega}_1)\phi^*_{n\ell m}(\underset{\sim}{r}_2) e^{ik_n r_2} , \qquad (20.5)$$

$$\Phi_3 \underset{r_2\to\infty}{\widetilde{\quad}} r_2^{-1} \sum_{n\ell m} \tilde{g}_{n\ell m}(\hat{\Omega}_2)\phi^*_{n\ell m}(\underset{\sim}{r}_1) e^{ik_n r_1} . \qquad (20.6)$$

In the variational principle the functional

$$I = \int \Psi(H-E)\Phi d\underset{\sim}{r}_1 d\underset{\sim}{r}_2 \qquad (20.7)$$

is used. If Φ is a solution of the Schrödinger equation, then $I = 0$, and the variation of I has the form

$$\delta I = \int \Psi(H-E)\delta\Phi d\underset{\sim}{r}_1 d\underset{\sim}{r}_2 . \qquad (20.8)$$

We shall assume that Ψ is a solution of the Schrödinger equation with boundary conditions (1.3) and (1.4), and that only the function Φ_3 is varied in Φ, while Φ_0, Φ_1, and Φ_2 preserve their exact asymptotic form. The integral (20.8) can be transformed to an integral over surfaces of the form (17.5) and (17.12), where now instead of Φ^* we have $\delta\Phi$. Using the asymptotic expressions (1.3), (1.4), (2.23), and (20.4)-(20.6), we find that the two terms in (17.5) cancel each other in all cases and the only nonzero contribution to (17.12) comes from the part of the asymptotic form (1.3) that contains the plane wave. With the aid of the asymptotic form (15.10) of a plane wave we obtain

$$\delta I = 2\pi\delta\tilde{f}_0(-\underset{\sim}{k}_0) \quad . \tag{20.9}$$

If we consider the functional in which Ψ and Φ are interchanged, then the nonzero contribution to (17.5) comes from the term containing the six-dimensional plane wave Φ_0, which gives

$$\delta\int \Phi(H-E)\Psi d\underset{\sim}{r}_1 d\underset{\sim}{r}_2 = (2\pi)^{5/2}\left(\frac{i}{\kappa}\right)^{3/2} \delta A(-\underset{\sim}{k}_1,-\underset{\sim}{k}_2) \ . \tag{20.10}$$

The role of Φ_1 and Φ_2 was considered by Merkur'ev [64], who showed that they do not contribute to (20.10).

No numerical calculations based on the variational relations (20.9) and (20.10) have yet been performed.

The amplitude $\tilde{f}$ corresponds to the process inverse to ionization. Assuming that in the integral I the functions Ψ and Φ have their exact values, for which $I = 0$, we transform the integral to the form (17.5) and (17.12) and evaluate it using the asymptotic expressions. We thus find

$$\tilde{f}_0(-\underset{\sim}{k}_0) = \left(\frac{2\pi i}{\kappa}\right)^{3/2} A(-\underset{\sim}{k}_1,-\underset{\sim}{k}_2) \quad , \tag{20.11}$$

which expresses the principle of detailed balance for the

processes of ionization and recombination. By changing the directions of the vectors $\underset{\sim}{k}_0$, $\underset{\sim}{k}_1$, and $\underset{\sim}{k}_2$ and using (21.4) with $\Delta = 0$ for short-range forces, we can rewrite this expression in the form

$$\tilde{f}(\underset{\sim}{k}_1, \underset{\sim}{k}_2 \rightarrow \underset{\sim}{k}_0, n_0) = (2\pi)^{3/2} f(\underset{\sim}{k}_0, n_0 \rightarrow \underset{\sim}{k}_1, \underset{\sim}{k}_2) \quad . \qquad (20.12)$$

Here the initial and final quantum numbers of the electrons are used as the arguments. The factor $(2\pi)^{3/2}$ stems from the fact that the initial-state plane waves in both (1.3) and (20.2) are unnormalized; moreover (1.3) contains one plane wave and (20.2) contains two.

CHAPTER VI

ELECTRON EXCHANGE IN IONIZATION

§21. *Phase Shift of the Scattered-Electron Wave Function*

In the asymptotic forms (1.3) and (1.4) we introduced the quantity η, to account for the phase shift arising from long-range Coulomb forces. In order to obtain an explicit expression for η, we must compare the asymptotic forms (2.20) and (2.23). The logarithmic phase shift γ was found in Chapter II. Using Eq. (9.11) for γ we obtain

$$\arg f(\underset{\sim}{v}_1,\underset{\sim}{v}_2) \quad \frac{3\pi}{4} + \frac{1}{v_2} \ell n \; 2v_2 r_2 + \eta(\underset{\sim}{v}_2,\underset{\sim}{r}_1) =$$

$$= \arg A(\hat{\Omega}) + \left(\frac{1}{v_1} + \frac{1}{v_2} - \frac{1}{v_{12}}\right) \ell n \; \kappa r \quad . \qquad (21.1)$$

Note that the choice of the form of γ plays no role here, because the right side of (21.1) is left unchanged by the transformation (9.3) and (9.4). We employ further the relations, which follow from (2.4) and (2.5),

$$r = \frac{\kappa}{v_1} r_1 \quad , \quad r_2 = \frac{v_2}{v_1} r_1 \quad . \qquad (21.2)$$

We may then write

$$\eta(\underset{\sim}{v}_2,\underset{\sim}{r}_1) = \left(\frac{1}{v_1} - \frac{1}{|\underset{\sim}{v}_1-\underset{\sim}{v}_2|}\right) \ell n \; v_1 r_1 + \beta(\underset{\sim}{v}_1,\underset{\sim}{v}_2) \quad , \qquad (21.3)$$

where the phase shift β is independent of r_1 but does depend on the choice for the phase of the amplitude f. In this sense the logarithmic phase shift η, analogous to γ, is not uniquely defined. If we consider the phase shift β to be given, then the relation between the amplitudes f and A takes the form

$$f(\underset{\sim}{v}_1,\underset{\sim}{v}_2) = e^{i\Delta(\underset{\sim}{v}_1,\underset{\sim}{v}_2)} \left(\frac{i}{\kappa}\right)^{3/2} A(\underset{\sim}{v}_1,\underset{\sim}{v}_2) \quad , \tag{21.4}$$

where

$$\Delta(\underset{\sim}{v}_1,\underset{\sim}{v}_2) = \beta_0(\underset{\sim}{v}_1,\underset{\sim}{v}_2) - \beta(\underset{\sim}{v}_1,\underset{\sim}{v}_2) \quad , \tag{21.5}$$

$$\beta_0(\underset{\sim}{v}_1,\underset{\sim}{v}_2) = \left(\frac{1}{v_1} - \frac{1}{|\underset{\sim}{v}_1-\underset{\sim}{v}_2|}\right) \ln\frac{\kappa^2}{v_1^2} + \frac{1}{v_2}\ln\frac{\kappa^2}{2v_2^2} \quad . \tag{21.6}$$

On the basis of the relations (2.4) and (2.5) we have used $\underset{\sim}{v}_1$ and $\underset{\sim}{v}_2$ instead of $\hat{\Omega}$ as the arguments of the amplitude A. Similarly, comparing (2.22) and (2.23), we obtain

$$g(\underset{\sim}{v}_1,\underset{\sim}{v}_2) = e^{i\Delta(\underset{\sim}{v}_1,\underset{\sim}{v}_2)} \left(\frac{i}{\kappa}\right)^{3/2} A(\underset{\sim}{v}_2,\underset{\sim}{v}_1) \quad . \tag{21.7}$$

The relation between the amplitudes f and g takes the form

$$g(\underset{\sim}{v}_1,\underset{\sim}{v}_2) = e^{i\tau(\underset{\sim}{v}_1,\underset{\sim}{v}_2)} f(\underset{\sim}{v}_2,\underset{\sim}{v}_1) \quad , \tag{21.8}$$

where

$$\tau(\underset{\sim}{v}_1,\underset{\sim}{v}_2) = \Delta(\underset{\sim}{v}_1,\underset{\sim}{v}_2) - \Delta(\underset{\sim}{v}_2,\underset{\sim}{v}_1) \quad . \tag{21.9}$$

If $\beta \neq \beta_0$, then for $\underset{\sim}{v}_1 \neq \underset{\sim}{v}_2$ the amplitudes f and g differ in phase. In approximate calculations, the wave function does not usually yield the asymptotic form (1.3); for example, in the Born approximation one obtains $\eta \equiv 0$. The phase Δ is then chosen by additional considerations (see §30).

The principal difference between the phase shifts η and γ is the absence of the second term from (9.11). This is to be expected because it is contained in the phase of the atomic wave function entering into (1.3). The expression for η can be elucidated also from a somewhat different point of view. A logarithmic phase shift of the form (2.13) corresponds to an equation of the Coulomb type

$$\left(\Delta_1 + \frac{2Z_1(\underset{\sim}{v}_2,\hat{\Omega}_1)}{r_1} + v_1^2\right) F(\underset{\sim}{v}_2,\underset{\sim}{r}_1) = 0 \quad , \tag{21.10}$$

where

$$Z_1 = 1 - \frac{v_1}{|\underset{\sim}{v}_1-\underset{\sim}{v}_2|} \quad . \tag{21.11}$$

The first term of the expression for Z_1 corresponds to the interaction between the first electron and the nucleus, and the second term to the interaction between the two electrons. Indeed, the potential energy of the first electron in the field of the nucleus and the second electron is

$$V_1 = -\frac{1}{r_1} + \frac{1}{|\underset{\sim}{r}_1-\underset{\sim}{r}_2|} \quad . \tag{21.12}$$

From (2.4) and (2.5) we obtain

$$|\underset{\sim}{r}_1-\underset{\sim}{r}_2| = \frac{|\underset{\sim}{v}_1-\underset{\sim}{v}_2|}{v_1} r_1 \quad , \tag{21.13}$$

which implies

$$V_1 = -Z_1/r_1 \quad . \tag{21.14}$$

We note that η and Z_1 depend on the velocities of the outgoing electrons and the angle between them. The charge Z_1 is almost zero if $v_2 << v_1$; in this case the slow electron screens the nucleus.

The Eqs. (21.10) and (21.11) also follow from the system of equations for the wave functions of the scattered electron. We look for a solution of the Schrödinger equation in the form

$$\Psi = \Psi_d + \int F(\underset{\sim}{v}_2,\underset{\sim}{r}_1)\phi(\underset{\sim}{v}_2,\underset{\sim}{r}_2)d\underset{\sim}{v}_2 \quad , \tag{21.15}$$

where Ψ_d describes the excitation of the discrete levels. According to (1.3) we have

$$F(\underset{\sim}{v}_2,\underset{\sim}{r}_1) \underset{r\to\infty}{\widetilde{\quad}} r_1^{-1} f(\underset{\sim}{v}_1,\underset{\sim}{v}_2)e^{iv_1r_1+i\eta(\underset{\sim}{v}_2,\underset{\sim}{r}_1)} \quad . \tag{21.16}$$

Substituting (21.15) into (1.1) we obtain

$$\left(\Delta_1 + \frac{2}{r_1} + v_1^2\right)F(\underset{\sim}{v}_2,\underset{\sim}{r}_1) = 2U(\underset{\sim}{v}_2,\underset{\sim}{r}_1) \quad , \tag{21.17}$$

where

$$U = U_d + \int Y(\underset{\sim}{v}_2,\underset{\sim}{v},\underset{\sim}{r}_1)F(\underset{\sim}{v},\underset{\sim}{r}_1)d\underset{\sim}{v} \quad , \tag{21.18}$$

$$Y = \int \phi^*(\underset{\sim}{v}_2,\underset{\sim}{r}_2)\frac{1}{r_{12}}\phi(\underset{\sim}{v},\underset{\sim}{r}_2)d\underset{\sim}{r}_2 \quad . \tag{21.19}$$

We are interested in the behavior of U for $r_1 \to \infty$, so U_d can be ignored. The integral for Y can be found if the Coulomb functions are replaced by normalized plane waves

$$\phi(\underset{\sim}{v},\underset{\sim}{r}) \to (2\pi)^{-3/2}\exp(i\underset{\sim}{v}\cdot\underset{\sim}{r}) \quad . \tag{21.20}$$

Then we obtain

$$Y = \frac{\exp[i(\underset{\sim}{v}-\underset{\sim}{v}_2)\cdot\underset{\sim}{r}_1]}{2\pi^2(\underset{\sim}{v}-\underset{\sim}{v}_2)^2} \quad . \tag{21.21}$$

This expression is singular for $\underset{\sim}{v} = \underset{\sim}{v}_2$. Since the singularity is due to the behavior of the integrand at large distances, the substitution (21.20) is justified in the neighborhood of the singular point. On the other hand, the asymptotic form of the potential is determined only by the neighborhood of the singular point.

We substitute (21.16) and (21.21) into (21.18), and note that now $v_1 = (2E-v^2)^{1/2}$. We then take outside of the integral the amplitude f and the exponential $e^{i\eta}$, using their values at the singular point. To the remaining integral we apply the formula [35]

$$\int \frac{\exp[it\Phi(\mathbf{r})]d\mathbf{r}}{|\mathbf{r}-\mathbf{r}_0|^2} \underset{t\to\infty}{\sim} \frac{2\pi^2\exp[it\Phi(\mathbf{r}_0)]}{t|\mathrm{grad}\Phi(\mathbf{r}_0)|} \quad . \tag{21.22}$$

The result can be written in the form [31]

$$U \sim \frac{v_1}{r_1|\mathbf{v}_1-\mathbf{v}_2|} F(\mathbf{v}_2,\mathbf{r}_1) \quad , \tag{21.23}$$

which implies (21.10) and (21.11).

§22. *Exchange Effects in Ionization of Hydrogen*

The electron-exchange effects, which occur in the ionization of the hydrogen atom, are also characteristic of other singly-valent atoms, when the incident and valence electrons can be considered to be moving in a given field of the nucleus and the remaining electrons.

The collision process with indistinguishable electrons is described by wave functions with symmetric and antisymmetric spatial parts

$$\Psi^{\pm} = \Psi(\mathbf{r}_1,\mathbf{r}_2) \pm \Psi(\mathbf{r}_2,\mathbf{r}_1) \quad . \tag{22.1}$$

The asymptotic form of the function $\Psi^{\pm}$ for $r_1 \to \infty$ is obtained by replacing the amplitude f in (1.3) by

$$f^{\pm} = f \pm g \quad . \tag{22.2}$$

Instead of (1.4) we obtain an expression which differs from the asymptotic form for $r_1 \to \infty$ by the transposition of $\underset{\sim}{r}_1$ and $\underset{\sim}{r}_2$ and by the ± sign. We note that now the incoming wave is present both as $r_1 \to \infty$ and as $r_2 \to \infty$.

From (21.8) and (22.2) we obtain

$$f^{\pm}(\underset{\sim}{v}_1, \underset{\sim}{v}_2) = e^{i\tau(\underset{\sim}{v}_1, \underset{\sim}{v}_2)} f^{\pm}(\underset{\sim}{v}_2, \underset{\sim}{v}_1) \quad . \tag{22.3}$$

Thus, on interchanging $\underset{\sim}{v}_1$ and $\underset{\sim}{v}_2$ the amplitude $f^{\pm}$ changes only its phase.

If the logarithmic phase shift γ is chosen to be symmetric with respect to interchange of $\underset{\sim}{r}_1$ and $\underset{\sim}{r}_2$, for example, in the form (9.11), then the function $\Psi^{\pm}$ has the asymptotic form (2.23), where A is replaced by

$$A^{\pm}(\hat{\Omega}) = A(\hat{\Omega}) \pm A(\hat{\hat{\Omega}}) \quad . \tag{22.4}$$

Here $\hat{\hat{\Omega}}$ differs from $\hat{\Omega}$ by the interchange of $\hat{\Omega}_1$ and $\hat{\Omega}_2$ and by the substitution of $\pi/2 - \alpha$ for α. Equation (21.4) holds for $A^{\pm}$ and $f^{\pm}$.

The electron fluxes satisfy the equations

$$\underset{\sim}{J}_1^{\pm}(\underset{\sim}{r}_1, \underset{\sim}{r}_2) = \underset{\sim}{J}_2^{\pm}(\underset{\sim}{r}_2, \underset{\sim}{r}_1) \quad , \tag{22.5}$$

$$\underset{\sim}{J}_1^{\pm}(\underset{\sim}{r}) = \underset{\sim}{J}_2^{\pm}(\underset{\sim}{r}) \quad . \tag{22.6}$$

Each electron flux now takes the asymptotic form (3.5) with f replaced by $f^{\pm}$. With indistinguishable electrons one measures experimentally the total flux of both electrons, which due to

(22.6) is equal to twice the flux of an individual electron. Thus, the incident-electron flux density is now equal to $2k_0$ [the factor 2 would be absent if we had included a factor $1/\sqrt{2}$ in the definition (22.1) of $\Psi^{\pm}$]. On the other hand, in defining the ionization cross section only the flux of one electron should be considered, i.e., only half the total flux of outgoing electrons. Thus, instead of (3.11) we obtain

$$\sigma^{\pm} = \frac{1}{2}\int_0^E \frac{v_1 v_2}{k_0}\, d\varepsilon_2 \int |f^{\pm}(\underset{\sim}{v}_1, \underset{\sim}{v}_2)|^2 d\hat{\Omega}_1 d\hat{\Omega}_2 \quad . \tag{22.7}$$

With the aid of (22.3) it is not difficult to see that the integrand in (22.7), integrated over $\hat{\Omega}_1$ and $\hat{\Omega}_2$, is symmetric with respect to the substitution of $\varepsilon_1 = E - \varepsilon_2$ for ε_2. Therefore, the factor 1/2 can be replaced by an integration to E/2:

$$\sigma^{\pm} = \int_0^{E/2} \frac{v_1 v_2}{k_0}\, d\varepsilon_2 \int |f(\underset{\sim}{v}_1, \underset{\sim}{v}_2) \pm g(\underset{\sim}{v}_1, \underset{\sim}{v}_2)|^2 d\hat{\Omega}_1 d\hat{\Omega}_2 \quad . \tag{22.8}$$

This expression differs from the cross section ignoring exchange (3.13) only in the interference term. It demonstrates that in the ionization of hydrogen the effect of exchange reduces to interference. The physical interpretation of the integrand in (22.8) is analogous to that in (3.13); they differ only in that (22.8) refers to indistinguishable electrons. When A is replaced by $A^{\pm}$ in Eq. (3.15), the integration over α is confined to the range $0 \le \alpha \le \pi/4$.

Using (21.8) we can rewrite Eq. (22.8) in the form

$$\sigma^{\pm} = \sigma_0 \pm 2\sigma_{int} \quad , \tag{22.9}$$

where σ_0 denotes the cross section ignoring exchange, which is given by (3.11)-(3.13), and σ_{int} is the interference term:

$$\sigma_{\text{int}} = \int_0^{E/2} \frac{v_1 v_2}{k_0} d\varepsilon_2 \int \text{Re}[f(\underset{\sim}{v}_1,\underset{\sim}{v}_2)e^{-i\tau(\underset{\sim}{v}_1,\underset{\sim}{v}_2)} f^*(\underset{\sim}{v}_2,\underset{\sim}{v}_1)]d\hat{\Omega}_1 d\hat{\Omega}_2 \,. \tag{22.10}$$

For the cross section averaged over spin

$$\sigma = \frac{1}{4}\sigma^+ + \frac{3}{4}\sigma^- \quad , \tag{22.11}$$

we find

$$\sigma = \sigma_0 - \sigma_{\text{int}} \quad . \tag{22.12}$$

If we use the wave function $\Psi^\pm$ in calculating the probability flux referred to a closed hypersurface, as in §4, we find that Eqs. (4.7)-(4.12) retain their form, with the replacement of f by $f^\pm$, and Eqs. (4.13)-(4.16) are unchanged. The optical theorem takes the form

$$\sigma_d^\pm + \sigma_{\text{ion}}^\pm = \frac{4\pi}{k_0} \text{Im}[f_0(0) \pm g_0(0)] \quad . \tag{22.13}$$

Comparison of (22.13) with (4.19) shows that the sum of the interference terms can be expressed in terms of the imaginary part of the forward, elastic-scattering, exchange amplitude:

$$\sigma_d^{\text{int}} + \sigma_{\text{ion}}^{\text{int}} = \frac{2\pi}{k_0} \text{Im}\, g_0(0) \quad . \tag{22.14}$$

The expansion (5.25) for the function $\Psi^\pm$ takes the form

$$\Psi^\pm = \sum_\beta [F_\beta^\pm(\underset{\sim}{r}_1)\phi_\beta(\underset{\sim}{r}_2) \pm F_\beta^\pm(\underset{\sim}{r}_2)\phi_\beta(\underset{\sim}{r}_1)] \quad , \tag{22.15}$$

where

$$F_\beta^\pm = F_\beta \pm G_\beta \quad . \tag{22.16}$$

The function $F_\beta^\pm$ is nonsingular and, in analogy to (5.29), is orthogonal to the atomic wave functions with energies less than ε_β. The boundary conditions (5.30) and (5.32) also hold for this function (with f replaced by $f^{\pm}$).

The functions $F^{\pm}_{\beta}$, as well as F_β and G_β, are determined by a system of integro-differential equations, which is obtained by substituting the expansion (5.22) or (22.15) into the Schrödinger equation and requiring that

$$\int \phi^*_\beta(\underset{\sim}{r}_2)(H - E)\Psi(\underset{\sim}{r}_1,\underset{\sim}{r}_2)d\underset{\sim}{r}_2 = 0 \quad . \tag{22.17}$$

The expansion (22.15) has the advantage that the system of equations for $F^{\pm}_{\beta}$ consists of two independent systems for F^{+}_{β} and F^{-}_{β}, while the equations for F_β and G_β are coupled.

§23. *Allowance for Exchange in the Ionization of the Helium Atom*

The ionization of the helium atom poses the problem of three electrons in the field of a fixed center (we consider the mass of the nucleus to be infinite). For the sake of simplicity we assume the potentials to be short-range ($1/r \to e^{-cr}/r$, $c > 0$). The Coulomb case differs mainly by additional phase factors. The logarithmic phase shift γ for an e-He collision can be easily obtained from the general equations given in §§7 and 9, and the phase shift η differs from γ by the absence of the logarithmic terms, which are contained in the atomic wave function.[1] In the frozen-core approximation the principal results found for helium pertain also to the ionization of other doubly-valent atoms.

---

[1] The ionization of helium with allowance for the Coulomb phase shifts was considered by Tweed [135].

In the following, we use atomic units. We shall first consider the electrons to be distinguishable and suppose that the incident electron has the index 1. The solution of the Schrödinger equation then has the asymptotic form

$$\Psi \underset{r_1\to\infty}{\sim} \delta_{nn_0} e^{i\mathbf{k}_0\cdot\mathbf{r}_1}\chi_{n_0}(\mathbf{r}_2,\mathbf{r}_3) + r_1^{-1}\sum_n f_n(\hat{\Omega}_1)\chi_n(\mathbf{r}_2,\mathbf{r}_3)e^{ik_n r_1} , \tag{23.1}$$

$$\Psi \underset{r_2\to\infty}{\sim} r_2^{-1}\sum_n g_n(\hat{\Omega}_2)\chi_n(\mathbf{r}_3,\mathbf{r}_1)e^{ik_n r_2} , \tag{23.2}$$

$$\Psi \underset{r_3\to\infty}{\sim} r_3^{-1}\sum_n w_n(\hat{\Omega}_3)\chi_n(\mathbf{r}_1,\mathbf{r}_2)e^{ik_n r_3} . \tag{23.3}$$

The symbol n denotes the set of quantum numbers of the helium atom, χ_n is an eigenfunction of the helium atom, and k_n is related to the energy of the atom E_n by

$$\frac{1}{2} k_n^2 + E_n = E . \tag{23.4}$$

The sum over n in Eqs. (23.1)-(23.3) also includes an integration over the continuous spectrum of the atom and is performed over the range $E_n \le E$. If n refers to the discrete spectrum, f_n is the amplitude for direct excitation of the atom, and g_n and w_n are the amplitudes for exchange excitation. Since short-range potentials are assumed, there is no logarithmic phase shift η.

The eigenfunctions of the discrete spectrum of the atom are necessarily either symmetric or antisymmetric with respect to interchange of the electrons. We assume that the eigenfunctions of the continuous spectrum of the atom are also chosen to be symmetric or antisymmetric. The continuous spectrum of helium is divided into two regions, which can be called singly and doubly continuous. In the first region one of the electrons is in a bound state of the He^+ ion, and in the second region both electrons have positive energy. In the first case the wave function describes the scattering by the He^+ ion, and in the second case it describes the scattering of two electrons by the He^{++} ion. We choose the wave functions of the continuous spectrum with the asymptotic form "incident wave + converging wave," i.e., in the asymptotic forms (1.3) and (20.3)-(20.6) we replace $e^{ik_n r}$ and $e^{i\kappa r}$ by $e^{-ik_n r}$ and $e^{-i\kappa r}$. Under these conditions f_n, g_n, and w_n represent scattering (ionization) amplitudes.

We can write the asymptotic form of a wave function in the continuous spectrum of helium as

$$\chi_n(\underset{\sim}{r}_2,\underset{\sim}{r}_3) \sim \frac{1}{\sqrt{2}}\,[\phi_\beta(\underset{\sim}{r}_2)\phi_\gamma(\underset{\sim}{r}_3) \pm \phi_\beta(\underset{\sim}{r}_3)\phi_\gamma(\underset{\sim}{r}_2)] + \text{converging wave}, \tag{23.5}$$

where ϕ_β is an eigenfunction of the He^+ ion, which in the continuum is chosen with an asymptotic form analogous to (5.3):

$$\phi_{\underset{\sim}{v}} \sim (2\pi)^{-3/2} e^{i\underset{\sim}{v}\cdot\underset{\sim}{r}} + \text{converging wave} \quad . \tag{23.6}$$

The symbols β and γ denote the set of quantum numbers of the He^+ ion, and n denotes the set β, γ, and in addition the sign "+" or "-." If E_n lies in the continuum,

$$E_n = \varepsilon_\beta + \varepsilon_\gamma \quad , \tag{23.7}$$

where ε_β and ε_γ are the energies of the He^+ ion. To form a complete system of symmetric and antisymmetric atomic wave functions, it is sufficient for the indices β and γ to take values in the domain for which

$$\varepsilon_\beta \geq \varepsilon_\gamma \quad . \tag{23.8}$$

This condition should be taken into account when summing over n in Eqs. (23.1)-(23.3).

The amplitude f_n can be represented as an integral

$$f_n(\mathbf{k}_n) = -\frac{1}{2\pi}\int \Psi(\mathbf{r}_1,\mathbf{r}_2,\mathbf{r}_3)(H-E)\Phi_n^*(\mathbf{k}_n;\mathbf{r}_1,\mathbf{r}_2,\mathbf{r}_3)d\mathbf{r}_1 d\mathbf{r}_2 d\mathbf{r}_3 \ , \tag{23.9}$$

where Φ_n describes the final state of the system

$$\Phi_n = e^{i\mathbf{k}_n\cdot\mathbf{r}_1}\chi_n(\mathbf{r}_2,\mathbf{r}_3) \quad . \tag{23.10}$$

Analogous expressions, differing only by a permutation of the electrons in the final-state wave function, hold for the amplitudes g_n and w_n.

To derive Eq. (23.9) we transform the integral to a sum of integrals analogous to (17.12),

$$I = I_1 + I_2 + I_3 \quad , \tag{23.11}$$

where

$$I_j = \frac{1}{2}\lim_{r_1\to\infty}\int\left[\Phi_n^*\frac{\partial\Psi}{\partial r_j} - \Psi\frac{\partial\Phi_n^*}{\partial r_j}\right]d\hat{\Omega}_j d\mathbf{r}_t d\mathbf{r}_s \quad . \tag{23.12}$$

We integrate over the angular coordinates of the *jth* electron and over all coordinates of the remaining two electrons. Substituting (23.1) and the asymptotic form (15.10) of a plane wave into (23.12) and taking into account the orthonormality of the atomic wave functions, we obtain

$$I_1 = -2\pi f_n(\mathbf{k}_n) \quad . \tag{23.13}$$

If χ_n belongs to the discrete spectrum, it falls off rapidly for r_2 or $r_3\to\infty$, so that

$$I_2 = 0 \quad , \quad I_3 = 0 \quad . \tag{23.14}$$

Equations (23.13) and (23.14) imply (23.9). If χ_n belongs to the continuous spectrum, then I_2 and I_3 do not have definite values in the limit of $r_j \to \infty$. As in §17 we shall assume that in this case I_2 and I_3 are understood in the regularized sense, e.g., in the sense of an average over large radius [52], which then leads to (23.14).

Only those terms of Φ_n which contain a diverging wave in the asymptotic region can contribute to I. From this requirement, together with Eqs. (23.5) and (23.6) it is not difficult to see that the magnitude of the integral is not changed if for the final-state wave function we use

$$\Phi_n = \frac{(2\pi)^{3/2}}{\sqrt{2}} \phi_\alpha(\underset{\sim}{r}_1)[\phi_\beta(\underset{\sim}{r}_2)\phi_\gamma(\underset{\sim}{r}_3) \pm \phi_\beta(\underset{\sim}{r}_3)\phi_\gamma(\underset{\sim}{r}_2)] \ , \tag{23.15}$$

where the index α denotes the vector $\underset{\sim}{k}_n$. Substituting (23.15) into (23.9), we obtain [20]

$$f_n(\underset{\sim}{k}_n) \equiv f^{\pm}(\alpha,\beta,\gamma) = \frac{1}{\sqrt{2}} [a(\alpha,\beta,\gamma) \pm a(\alpha,\gamma,\beta)] \ , \tag{23.16}$$

where

$$a(\alpha,\beta,\gamma) = -\sqrt{2\pi}\int \Psi(\underset{\sim}{r}_1,\underset{\sim}{r}_2,\underset{\sim}{r}_3)(H-E)\phi^*_\alpha(\underset{\sim}{r}_1)\phi^*_\beta(\underset{\sim}{r}_2)\phi^*_\gamma(\underset{\sim}{r}_3)d\underset{\sim}{r}_1 d\underset{\sim}{r}_2 d\underset{\sim}{r}_3 \ . \tag{23.17}$$

Similarly we find

$$g_n(\underset{\sim}{k}_n) \equiv g^{\pm}(\alpha,\beta,\gamma) = \frac{1}{\sqrt{2}} [a(\gamma,\alpha,\beta) \pm a(\beta,\alpha,\gamma)] \ , \tag{23.18}$$

$$w_n(\underset{\sim}{k}_n) \equiv w^{\pm}(\alpha,\beta,\gamma) = \frac{1}{\sqrt{2}} [a(\beta,\gamma,\alpha) \pm a(\gamma,\beta,\alpha)] \ . \tag{23.19}$$

Equations (23.16), (23.18), and (23.19), as well as (21.4), (21.7), and (21.8), can be obtained by comparing the asymptotic forms (23.1)-(23.3) among themselves when two or three radius vectors are large simultaneously. This method of obtaining these equations does not require any regularization of the integrals.

The solution of the Schrödinger equation determined by the boundary conditions (23.1)-(23.3) has definite symmetry properties under interchange of the second and third electrons:

$$\Psi(1,2,3) = \pm\Psi(1,3,2) \ , \text{ if } \ \chi_0(2,3) = \pm\chi_0(3,2) \quad , \qquad (23.20)$$

where χ_0 describes the initial state of the atom. The numbers 1, 2, 3 designate $\underset{\sim}{r}_1$, $\underset{\sim}{r}_2$, $\underset{\sim}{r}_3$ respectively. To prove (23.20), we note that Ψ can be represented as a sum

$$\Psi = \frac{1}{2}\,(\Psi^+ + \Psi^-) \ , \ \Psi^\pm = \Psi(1,2,3) \pm \Psi(1,3,2) \quad . \qquad (23.21)$$

Owing to the invariance of the Hamiltonian under permutation of the electrons, the functions Ψ^+ and Ψ^- also satisfy the Schrödinger equation. However, only one of these contains the incoming wave in its asymptotic form. Therefore, the other function is equal to zero, because the conservation of probability excludes the existence of a regular solution containing only diverging waves.

Equations (23.17) and (23.20) imply

$$a(\alpha,\beta,\gamma) = \pm a(\alpha,\gamma,\beta) \quad . \qquad (23.22)$$

For the singlet initial state we have

$$f_n^- = 0 \quad , \quad g_n^\pm = \pm w_n^\pm \quad , \qquad (23.23)$$

and for the triplet initial state we find

$$f_n^+ = 0 \quad , \quad g_n^\pm = \mp w_n^\pm \quad . \qquad (23.24)$$

The signs "+" and "-" refer to the symmetry type of the index n. These relations indicate that states of the other symmetry type can be excited only by electron exchange.

The coordinate-space wave functions with definite symmetry properties corresponding to a definite value of the total electron spin have physical significance [1,73]. For the singlet initial state the total spin is S = 1/2. Equation (23.20) then implies that the function

$$\Psi_1 = \Psi(1,2,3) - \Psi(2,3,1) \tag{23.25}$$

has the appropriate symmetry properties. We find the asymptotic form of this function by making the following substitutions in (23.1)-(23.3),

$$f_n \to f_n - w_n \quad , \quad g_n \to g_n - f_n \quad , \quad w_n \to w_n - g_n \quad . \tag{23.26}$$

In addition, the incoming wave is now present also for $r_2 \to \infty$. The overall flux of incoming electrons is equal to $2k_0$. The flux of the *ith* electron, as in (3.3) and (3.4), is given by

$$\mathbf{j}_i = \int \mathrm{Im}(\Psi^* \nabla_i \Psi) d\mathbf{r}_t d\mathbf{r}_s \quad , \tag{23.27}$$

where the integration is performed over the coordinates of the two remaining electrons. Using the asymptotic form of the function Ψ_1 and considering the total flux of all three electrons, we obtain the following expressions for the differential cross sections for excitation of singlet and triplet states of the n*th* level:

$$\sigma_n^+ = \frac{k_n}{k_0} |f_n^+ - g_n^+|^2 \quad , \quad \sigma_n^- = 3 \frac{k_n}{k_0} |g_n^-|^2 \quad . \tag{23.28}$$

If the initial state of the atom is a triplet state, then in light of (23.20) the following coordinate-space wave functions are physically significant:

$$\Psi_2 = \Psi(1,2,3) + \Psi(2,3,1) - 2\Psi(3,1,2) \quad , \tag{23.29}$$

$$\Psi_3 = \Psi(1,2,3) + \Psi(2,3,1) + \Psi(3,1,2) \quad . \tag{23.30}$$

The function Ψ_2 corresponds to spin $S = 1/2$ and Ψ_3 to spin $S = 3/2$. Comparing the incident and scattered fluxes and taking into account the statistical weights of states with different total spins, we obtain in this case

$$\sigma_n^+ = \frac{k_n}{k_0} |g_n^+|^2 \ , \ \sigma_n^- = \frac{k_n}{k_0} \left(|f_n^- + g_n^-|^2 + 2|g_n^-|^2\right) \quad . \tag{23.31}$$

Substituting Eqs. (23.16), (23.18), and (23.19) into Eqs. (23.28) and (23.31) and considering (23.22), we obtain

$$\sigma_{\beta\gamma} \equiv \sigma_{\beta\gamma}^+ + \sigma_{\beta\gamma}^- = 2 \frac{k_n}{k_0} q(\alpha,\beta,\gamma) \quad , \tag{23.32}$$

where

$$q = |a_1|^2 + |a_2|^3 + |a_3|^2 \mp \mathrm{Re}(a_1 a_2^* + a_1 a_3^* + a_2 a_3^*) \ , \tag{23.33}$$

$$a_1 = a(\alpha,\beta,\gamma) \ , \ a_2 = a(\beta,\gamma,\alpha) \ , \ a_3 = a(\gamma,\alpha,\beta) \quad . \tag{23.34}$$

The minus sign in (23.33) is to be taken if the initial state of the atom is a singlet and the plus sign if it is a triplet. Considering (23.22) we see that $q(\alpha,\beta,\gamma)$ is invariant under permutations of the arguments.

For single ionization of helium, we have $\varepsilon_\gamma < 0$ by virtue of the condition (23.8). By integrating over $\beta \equiv \underset{\sim}{v}_\beta$ in the domain for which $0 \le \varepsilon_\beta \le E - \varepsilon_\gamma$, we find the cross section for ionization with the ion left in the state γ. Since two electrons leave in single ionization, the result should be divided by 2. Considering the invariance of q, division by 2 may be replaced by integration over the domain for which

$$0 \le \varepsilon_\beta \le \varepsilon_{\beta m} \ , \ \varepsilon_{\beta m} = \frac{1}{2}(E - \varepsilon_\gamma) \quad . \tag{23.35}$$

We then obtain

$$\sigma_\gamma = 2 \int_0^{\varepsilon_{\beta m}} \frac{v_\alpha v_\beta}{k_0} d\varepsilon_\beta \int q(\underset{\sim}{v}_\alpha, \underset{\sim}{v}_\beta, \gamma) d\hat{\Omega}_\alpha d\hat{\Omega}_\beta \quad . \qquad (23.36)$$

This result differs from the analogous equation (22.8) for the ionization of a hydrogen atom by the factor 2, reflecting the presence of two valence electrons, and also by the presence of the amplitude a_3, which represents a transition of the incoming electron to a bound state. This exchange process does not reduce to interference.

The domain of integration for the double ionization cross section is given by the conditions

$$0 \le \varepsilon_\gamma \le \varepsilon_\beta \; , \; \varepsilon_\beta + \varepsilon_\gamma \le E \quad . \qquad (23.37)$$

The result here should be divided by 3, because three electrons escape, but instead of this we may reduce the domain of integration by this factor, by imposing the conditions

$$0 \le \varepsilon_\gamma \le \varepsilon_\beta \le \varepsilon_\alpha \le E \; , \; \varepsilon_\alpha + \varepsilon_\beta + \varepsilon_\gamma = E \quad . \qquad (23.38)$$

Then we have

$$\sigma = 2 \int \frac{v_\alpha v_\beta v_\gamma}{k_0} d\varepsilon_\beta d\varepsilon_\gamma \int q(\underset{\sim}{v}_\alpha, \underset{\sim}{v}_\beta, \underset{\sim}{v}_\gamma) d\hat{\Omega}_\alpha d\hat{\Omega}_\beta d\hat{\Omega}_\gamma \quad . \qquad (23.39)$$

We obtain the cross section for double ionization ignoring exchange if in (23.39) we omit the factor of 2, set $q = |a_1|^2$, and integrate over the six domains differing by permutations of the indices α,β,γ in (23.38). By virtue of the symmetry properties of Eq. (23.33) the result will differ only in the absence of the interference terms. Thus, in the double ionization of helium the exchange effects reduce to interference.

The representation of the wave function in the form

$$\Psi = \sum_n [F_n(1)\chi_n(2,3) + G_n(2)\chi_n(3,1) + W_n(3)\chi_n(1,2)] \quad (23.40)$$

corresponds to the boundary conditions (23.1)-(23.3). The functions F_n, G_n, and W_n are not uniquely defined. In order to obtain a clearer picture of the possible choices for these functions, we consider the simplified case, in which the atomic wave functions are of the form

$$\chi_n(2,3) = \frac{1}{\sqrt{2}} [\phi_\beta(2)\phi_\gamma(3) \pm \phi_\beta(3)\phi_\gamma(2)] \quad . \quad (23.41)$$

Expand Ψ in a series:

$$\Psi = \sum_{\alpha\beta\gamma} C_{\alpha\beta\gamma}\phi_\alpha(1)\phi_\beta(2)\phi_\gamma(3) \quad . \quad (23.42)$$

This series divides symmetrically into three parts corresponding to (23.40) if the *ith* part contains all terms from (23.42) in which the energy of the *ith* electron is greater than or equal to the energy of each of the two remaining electrons. Thus we find

$$F^{\pm}_{\beta\gamma} = \frac{1}{\sqrt{2}} \sum_{\varepsilon_\alpha \geq \varepsilon_\beta} (C_{\alpha\beta\gamma} \pm C_{\alpha\gamma\beta})\phi_\alpha \quad . \quad (23.43)$$

Similar expressions, differing by the substitution of $C_{\gamma\alpha\beta}$ and $C_{\beta\gamma\alpha}$ for $C_{\alpha\beta\gamma}$, hold for $G^{\pm}_{\beta\gamma}$ and $W^{\pm}_{\beta\gamma}$. These functions are non-singular since they are orthogonal to ϕ_α if $\varepsilon_\alpha < \varepsilon_\beta$.

Only terms satisfying the conservation of energy can contribute to the effective cross sections. In addition, we have the summation condition with respect to α in (23.43) and the condition (23.8), which taken together imply

$$\varepsilon_\gamma \leq \varepsilon_\beta \leq \varepsilon_\alpha \; , \; \varepsilon_\alpha + \varepsilon_\beta + \varepsilon_\gamma = E \quad . \quad (23.44)$$

If these conditions are violated, F_n, G_n, and W_n fall off faster than r^{-1} and do not contribute to the effective cross sections. For a given ε_γ Eq. (23.44) implies Eq. (23.35) (if $\varepsilon_\beta \geq 0$), and if $\varepsilon_\gamma \geq 0$, we get (23.38).

In going from Ψ to the functions (23.25), (23.29), and (23.30), the unknown functions F_n, G_n, and W_n are replaced by linear combinations, e.g., of the form (23.26); from the mathematical point of view these linear combinations are convenient because the system of equations splits into independent subsystems.

CHAPTER VII

THRESHOLD BEHAVIOR OF THE IONIZATION CROSS SECTION

§24. *Threshold Behavior of the Ionization Cross Section in Classical Mechanics*

In §18 we discussed the determination of the threshold behavior of the ionization amplitude by taking the limit $E \to 0$ in the integral representation. Another well-known method involves the following ideas. The configuration space is divided into two regions -- internal and external. The boundary between them is chosen such that the potential in the external region admits a fairly exact solution of the Schrödinger equation in analytical form. This boundary we denote by r_0, which may depend on the direction $\hat{\Omega}$. This solution (or linear combination of solutions) is matched at $r = r_0$ with the wave function Ψ in the internal region. In general Ψ is unknown, but as in §18 we can assume that Ψ and its derivatives go to a finite limit as $E \to 0$. Then from the matching conditions and the known properties of the solution in the external region we can find the energy dependence of the ionization amplitude as $E \to 0$. We note that here, as in §18, we are concerned with the form of the dependence and not with its absolute value. The quantity r_0 can be chosen arbitrarily large, but the greater it is, the smaller the interval of E in which we can consider the wave function $\Psi(r_0)$ to be a continuous function of E.

The threshold behavior of the cross section for single ionization of atoms and ions by electrons was determined by Wannier [50] using a similar method. Wannier made use of the circumstance that at large distances the Coulomb potential varies sufficiently slowly that for sufficiently large r_0 classical mechanics is valid in the external region. In contrast with the above method, Wannier examined not the behavior of the wave function in the external region but the behavior of the classical trajectories. Furthermore he assumed that the distribution of initial conditions for the classical trajectories at $r = r_0$ (i.e., the density of representative points in phase space) remains finite as $E \to 0$. This is sufficient to determine the threshold dependence of the ionization cross section. In §27 it is shown that the semiclassical approximation of quantum mechanics also leads to Wannier's result.

Wannier divides the external region into two zones: a free-particle zone and a Coulomb zone. As $E \to 0$, the free-particle zone recedes and the Coulomb zone grows without bound. The threshold behavior of the ionization cross section is determined by the properties of the motion in the Coulomb zone.

Wannier [50] noted and made use of two important properties of the ionization process at low energies: (i) If the two electrons escape from the nucleus with low velocities, then their Coulomb repulsion should have a considerable effect on their motion. This changes their trajectories so that in the end they depart in nearly opposite directions. Hence, at energies near the ionization threshold the differential ionization cross section should have a sharp maximum at $\theta = \pi$. (ii) At low energies ionization is produced only by those classical trajectories for which, at the inner boundary of the Coulomb zone, the two electrons are at nearly equal distances from the nucleus, i.e., $\alpha \approx \pi/4$.

Clearly, these properties follow from the form of the potential energy, which for the case of the ionization of an atom or ion can be written as in (6.1), where

$$Z(\alpha,\theta) = \frac{z}{\cos\alpha} + \frac{z}{\sin\alpha} - \frac{1}{\sqrt{1 - \cos\theta\sin 2\alpha}} , \qquad (24.1)$$

and z is the ion charge after ionization. The potential energy as a function of θ has a minimum at $\theta = \pi$ and at this value of θ as a function of α has a maximum at $\alpha = \pi/4$. Therefore, the trajectories approach $\theta = \pi$ and move away from $\alpha = \pi/4$. The point where $\alpha = \pi/4$ and $\theta = \pi$ is a stationary (saddle) point of the function $Z(\alpha,\theta)$. Note that $\theta = \pi$ means that the electrons are found on opposite sides of the nucleus on a straight line through it, and $\alpha = \pi/4$ means that the electrons are equidistant from the nucleus.

Let us examine the case in which the total orbital angular momentum of the electrons is zero, L = 0. The result applies also to L > 0 although in a smaller energy interval, because for L > 0 it is necessary to increase r_0 in order to be able to ignore the centrifugal energy. For L = 0 the trajectories depend only on the variables r, α, θ. The equations of motion are of the form

$$\frac{d^2r}{dt^2} = r\left(\frac{d\alpha}{dt}\right)^2 + \frac{1}{4} r \sin^2 2\alpha \left(\frac{d\theta}{dt}\right)^2 - \frac{Z}{r^2} , \qquad (24.2)$$

$$\frac{d}{dt}\left(r^2 \frac{d\alpha}{dt}\right) = \frac{1}{2} r^2 \sin 2\alpha \cos 2\alpha \left(\frac{d\theta}{dt}\right)^2 + \frac{1}{r}\cdot\frac{\partial Z}{\partial\alpha} , \qquad (24.3)$$

$$\frac{d}{dt}\left(r^2 \sin^2 2\alpha \frac{d\theta}{dt}\right) = \frac{4}{r}\cdot\frac{\partial Z}{\partial\theta} , \qquad (24.4)$$

where t is the time. The energy of the system has the form

$$E = \frac{1}{2}\left(\frac{dr}{dt}\right)^2 + \frac{1}{2}r^2\left(\frac{d\alpha}{dt}\right)^2 + \frac{1}{8}r^2 \sin^2 2\alpha \left(\frac{d\theta}{dt}\right)^2 - \frac{Z}{r} \quad . \qquad (24.5)$$

We shall examine the trajectories in the neighborhood of the stationary point. Expand $Z(\alpha,\theta)$ in a series

$$Z(\alpha,\theta) = Z_0 + \frac{1}{2}Z_1(\Delta\alpha)^2 + \frac{1}{8}Z_2(\Delta\theta)^2 + \ldots \quad , \qquad (24.6)$$

where

$$\Delta\alpha = \alpha - \pi/4 \ , \ \Delta\theta = \theta - \pi \quad . \qquad (24.7)$$

The factor 1/8 before Z_2 is introduced to simplify the later equations. From (24.1) it follows that

$$Z_0 = \frac{4z-1}{\sqrt{2}} \ , \quad Z_1 = \frac{12z-1}{\sqrt{2}} \ , \quad Z_2 = -\frac{1}{\sqrt{2}} \quad . \qquad (24.8)$$

Assume that $\Delta\alpha$ and $\Delta\theta$ as well as $d\alpha/dt$ and $d\theta/dt$ are small quantities. The assumption that the derivatives are small is justified if the directions of the trajectories are almost radial. We seek a solution in the form

$$\Delta\alpha = u_1(r) \ , \quad \Delta\theta = u_2(r) \quad . \qquad (24.9)$$

Substituting

$$\alpha = \pi/4 + u_1 \ , \quad \theta = \pi + u_2 \qquad (24.10)$$

into Eqs. (24.2)-(24.4) and keeping terms of equal order on both sides of the equations, we obtain

$$\frac{d^2r}{dt^2} = -\frac{Z_0}{r^2} \quad , \qquad (24.11)$$

$$\frac{d}{dt}\left(r^2\frac{du_i}{dt}\right) = \frac{Z_i u_i}{r} \quad , \quad i = 1,2 \quad . \qquad (24.12)$$

The equations for u_1 and u_2 are linear and uncoupled, which considerably simplifies the problem. From the energy expression (24.5) it follows that

$$\frac{dr}{dt} = \sqrt{2E + (2Z_0/r)} \equiv w(r) \quad , \tag{24.13}$$

which satisfies (24.11). By means of (24.13) we transform to differentiation with respect to r in Eq. (24.12). We then find

$$2r^2(Er + Z_0)\frac{d^2u_i}{dr^2} + r(4Er + 3Z_0)\frac{du_i}{dr} = Z_i u_i \quad . \tag{24.14}$$

The solutions are expressed in terms of the hypergeometric function:

$$u_i = C_{i1}u_{i1} + C_{i2}u_{i2} \quad , \tag{24.15}$$

$$u_{ij} = r^{m_{ij}}\,{}_2F_1\left(m_{ij}, m_{ij}+1, 2m_{ij} + \frac{3}{2}, -\frac{Er}{Z_0}\right) \quad , \tag{24.16}$$

$$m_{i1} = -\frac{1}{4} - \frac{1}{2}\mu_i, \quad m_{i2} = -\frac{1}{4} + \frac{1}{2}\mu_i \quad , \tag{24.17}$$

$$\mu_i = \frac{1}{2}\sqrt{1 + 8Z_i/Z_0} \quad . \tag{24.18}$$

The C_{ij} are arbitrary constants. Note that the parameters of the hypergeometric function in (24.16) are such that it reduces to a Legendre function.

Equations (24.8) and (24.18) imply

$$\mu_1 = \frac{1}{2}\sqrt{\frac{100z-9}{4z-1}} \quad , \quad \mu_2 = \frac{1}{2}\sqrt{\frac{4z-9}{4z-1}} \quad . \tag{24.19}$$

The functions μ_1 and μ_2 have a singularity for $z = 1/4$. If $z > 1/4$, then μ_1 is real and greater than 5/2. The function μ_2 is real, and less than 1/2, for $z > 9/4$; in the interval $1/4 < z < 9/4$ μ_2 is imaginary. For $z \to \infty$

$$\mu_1 \to \frac{5}{2} \quad , \quad \mu_2 \to \frac{1}{2} \quad . \tag{24.20}$$

If μ_2 is imaginary, then $u_{21} = u_{22}^*$. In this case u_2 is real if $C_{21} = C_{22}^*$.

Equation (24.16) takes simple forms in the limiting cases. If $Er << Z_0$, then

$$u_{ij} \sim r^{m_{ij}} \quad . \tag{24.21}$$

If $Er >> Z_0$, then by analytic continuation of the hypergeometric function [26] we obtain

$$u_{ij} \sim \gamma_{ij} \left[1 - \frac{Z_i}{2Er}\left(\ln \frac{Er}{Z_0} + h_{ij}\right)\right] \quad , \tag{24.22}$$

where

$$\gamma_{ij} = \left(\frac{Z_0}{E}\right)^{m_{ij}} \frac{\Gamma(2m_{ij}+3/2)}{\Gamma(m_{ij}+3/2)\Gamma(m_{ij}+1)} \quad , \tag{24.23}$$

$$h_{ij} = \psi(1) + \psi(2) - \psi\left(m_{ij} + \frac{1}{2}\right) - \psi(m_{ij}+1) \quad . \tag{24.24}$$

Here $\Gamma(x)$ denotes the gamma function, and $\psi(x)$ is the logarithmic derivative of the gamma function. We note the relation, following from the properties of the Γ and ψ functions,

$$\gamma_{i1}\gamma_{i2}(h_{i2} - h_{i1}) = 2\sqrt{Z_0 E}\,\frac{\mu_i}{Z_i} \quad . \tag{24.25}$$

Let us consider the behavior of the trajectories for $E \to 0$. First we investigate the dependence of $\Delta\theta$ on r and E. If the energy is sufficiently small, then Eq. (24.21) is applicable at the inner boundary of the Coulomb zone. For $z > 9/4$ we have $m_{2i} < 0$, and for $9/4 > z > 1/4$ we have $\mathrm{Re}\, m_{2i} < 0$. Hence as r increases, $\Delta\theta$ falls off monotonically or oscillates with decreasing amplitude. The region in which this occurs increases as E decreases. From the other asymptotic equation (24.22) it follows that as $r \to \infty$, $\Delta\theta$ goes to the limiting value $\Delta\theta(\infty) = C_{21}\gamma_{21} + C_{22}\gamma_{22}$. Since $|\gamma_{ij}| \sim E^{-\mathrm{Re}\, m_{ij}}$, for given C_{21} and C_{22} the limiting values decrease as $E \to 0$. This is in agreement with the behavior of the trajectories at the beginning of the Coulomb zone and reaffirms the first property mentioned earlier of the ionization process at low energies. For z = 1 and z = 2 we have

$$\Delta\theta(\infty) \sim E^{1/4} \quad . \tag{24.26}$$

The concentration of trajectories about $\theta = \pi$ does not influence the total ionization cross section but produces a maximum in the differential cross section, the width of which grows narrower in proportion to $E^{1/4}$. This behavior of the differential cross section was found by Vinkaln and Gailitis [74].

The threshold behavior of the total ionization cross section is determined by the dependence of $\Delta\alpha$ on r and E. If $z > 1/4$, then $m_{11} < -3/2$ and $m_{12} > 1$. At the inner boundary of the Coulomb zone u_{11} falls off as r increases and u_{12} grows. In agreement with this $u_{11}(\infty)$ falls off as $E \to 0$ and $u_{12}(\infty)$ grows. In order that $\Delta\alpha(\infty)$ remain small, it is necessary for C_{12} to decrease; we thus assume

$$C_{12} \sim E^{m_{12}} \quad . \tag{24.27}$$

If C_{12} is sufficiently large, then α deviates from $\pi/4$ as r increases. In the process the trajectory can leave the domain of validity of Eqs. (24.11) and (24.12). However, since the potential at $\theta = \pi$ continues to drop off monotonically as a function of α as $|\Delta\alpha|$ increases (see §14), we may conclude that $|\Delta\alpha|$ will continue to increase as r becomes larger. The trajectories will become more concentrated towards $\alpha = 0$ and $\alpha = \pi/2$. For these values of α a single electron emerges. Ionization occurs only for those trajectories for which α does not reach 0 or $\pi/2$ as $r \to \infty$. Hence, for ionizing trajectories the coefficient C_{12} must lie in some small interval $|C_{12}| \leq C_{max}$, where $C_{max} \to 0$ when $E \to 0$. Taking the similarity principle into consideration, we may assume that C_{max} falls off according to Eq. (24.27). The coefficients C_{ij} determine the initial conditions of the trajectories at the inner boundary of the Coulomb zone. If the distribution of initial conditions is finite in the neighborhood of $C_{12} \approx 0$, then the total ionization cross section varies in the same way as C_{max}; hence,

$$\sigma \sim E^{m_{12}} \quad . \tag{24.28}$$

The similarity principle which was employed above follows from the homogeneity of the Coulomb potential and is expressed by the invariant form of the trajectories under the transformation [75]

$$r \to ar\,, \quad \alpha \to \alpha\,, \quad \theta \to \theta\,, \quad t \to a^{3/2}t\,, \quad E \to a^{-1}E \quad . \tag{24.29}$$

From (24.29) it follows that in the equations determining the trajectories, α and θ depend on E and r through the product Er. In Eqs. (24.15) and (24.16) this is satisfied if we set $C_{ij} = D_{ij}E^{m_{ij}}$. The values of D_{ij} are the same for similar trajectories at different E. Equation (24.27) then follows.

We note that under the transformation (24.29) the effective cross section is multiplied by a^2. If $\sigma(\varepsilon_0, v_0)$ is the cross section for ionization (in classical mechanics) of an atom or ion from level ε_0 by an electron incident with velocity v_0, then

$$\sigma\left(\frac{\varepsilon_0}{a}\,,\ \frac{v_0}{\sqrt{a}}\right) = a^2\sigma(\varepsilon_0, v_0) \quad . \tag{24.30}$$

For the ionization of an atom (z = 1) Eqs. (24.17) and (24.19) yield

$$\sigma \sim E^{1.127} \quad , \tag{24.31}$$

and for the ionization of an ion (z = 2) they yield

$$\sigma \sim E^{1.056} \quad . \tag{24.32}$$

The threshold dependence differs slightly from a linear law and quickly approaches linearity as the ion charge is increased. According to §18 the threshold dependence is linear if the interaction between the electrons is short range, i.e., if it may be neglected in the Coulomb zone. In any actual problem the role of the interaction of the electrons is small because the electrons leave the nucleus in opposite directions.

If the potential in the Coulomb zone is of the form

$$V = -\frac{z}{r_1} - \frac{z}{r_2} \quad , \tag{24.33}$$

where there is no interaction between the electrons, we find

$$Z_0 = 2z\sqrt{2} \ , \quad Z_1 = 6z\sqrt{2} \ , \quad Z_2 = 0 \ , \quad m_{12} = 1 \quad , \tag{24.34}$$

which corresponds to a linear threshold dependence. Thus, the method of Wannier leads to the correct result in this limiting case. In the problem with the potential (24.33) there is no concentration of trajectories at $\theta = \pi$, so the dependence on α may be analyzed at any angle θ. Even if the electrons do not interact with one another, the trajectories nevertheless deviate from $\alpha = \pi/4$. In this case there is a simple interpretation. For electrons which move independently the energy of each is conserved,

$$\varepsilon_i = \frac{1}{2}\left(\frac{dr_i}{dt}\right)^2 + \frac{1}{2}\cdot\frac{\ell_i^2}{r_i^2} - \frac{z}{r_i} \quad , \tag{24.35}$$

where ℓ_i is the orbital angular momentum. The parameter C_{12} determines the difference $\varepsilon_1 - \varepsilon_2$. If $C_{12} \neq 0$ and at the beginning of the motion $r_1/r_2 \approx 1$, this ratio will change as r increases, and approach a limiting value of $\sqrt{\varepsilon_1/\varepsilon_2}$. The disappearance of a range of values of C_{12} that leads to ionization as $E \to 0$ results from the impossibility of ε_1 and ε_2 being simultaneously positive.

The threshold behavior of the N-fold ionization cross section can be described approximately by the equation due to Wannier [76],

$$\sigma_N \sim E^N \quad , \tag{24.36}$$

which is exact in the absence of the Coulomb interaction between electrons. For N = 2 this follows from the equations of §23, if we take the limit $E \to 0$ under the integral sign of (23.17) and recall that the Coulomb wave function $\phi_{\underset{\sim}{v}}(\underset{\sim}{r})$ increases like $v^{-1/2}$ as $v \to 0$. By analogy with the single-ionization case we can expect that for multiple ionization the effects of any interaction among the electrons will be small. Then Eq. (24.36) is nearly exact.

The approximation described by Eqs. (24.11) and (24.12) is inapplicable for large $\Delta\alpha$ and therefore gives no information about the distribution of trajectories over α at the outer boundary of the Coulomb zone. The distribution over α determines the energy distribution of electrons because as $r \to \infty$ the magnitude of α determines the ratio of the electron velocities. To study the energy distribution, Vinkaln and Gailitis [74] solved numerically the classical equations of motion on a computer for the simplified problem in which the electrons move along a straight line in opposite directions (i.e., $\theta = \pi$). The form of the initial conditions was analogous to (24.15) and (24.21), where $C_{21} = C_{22} = 0$. The calculations were performed for $C_{11} = 0$ and various C_{12}. As $r \to \infty$ the electron energies ε_1 and ε_2 as functions of C_{12} turned out to be practically straight lines. Owing to the smallness of the range of values of C_{12} leading to ionization, for small E all values of C_{12} in this range can be considered to be equally probable. Then

$$\frac{d\sigma}{d\varepsilon_1} \sim \frac{dC_{12}}{d\varepsilon_1} \sim \text{const.} \tag{24.37}$$

Thus, the differential cross section for small E is independent of ε_1, but it may depend on E. When σ depends on E as in (24.28), we obtain

$$\frac{d\sigma}{d\varepsilon_1} \sim E^{m_{12}-1} \quad . \tag{24.38}$$

According to (24.37) all possible energies ε_1 are equally probable. We note that the condition $\alpha \approx \pi/4$, which would mean $\varepsilon_1 \approx \varepsilon_2 \approx E/2$, applies not to large r but to the inner edge of the Coulomb zone. Using the fact that $\varepsilon_1 = E\cos^2\alpha$ for $r \to \infty$, we obtain

$$\frac{d\sigma}{d\alpha} \sim E^{m_{12}} \sin 2\alpha \quad . \tag{24.39}$$

Equation (24.38) can be extended to highly excited discrete states. For the cross section for excitation of the *nth* level we obtain

$$\sigma_n \sim n^{-3}|E|^{m_{12}-1} \quad . \tag{24.40}$$

The question of the correctness of the threshold formula (24.28) really involves two distinct questions. First, how accurate is this formula in classical mechanics? Second, how valid is classical mechanics in the present case?

The ionization cross section for the hydrogen atom has been calculated by Abrines *et al.* [77] and Bratsev and Ochkur [78] using the numerical methods of classical mechanics (computer integration of the trajectory equations). These authors do not discuss the threshold behavior of the ionization cross section. In the calculations it is practically impossible to examine the threshold behavior for all initial conditions. The problem is that to determine the ionization cross section it is necessary to discover the volumes of those regions in the space of initial conditions which lead to ionization trajectories. Owing to the multidimensionality of the space of initial conditions the volumes of the regions corresponding to ionization are determined with considerable absolute error. As $E \to 0$ the size of the ionization regions shrinks to zero and the relative error grows.

However, the problem can be simplified. As we saw earlier, among the coefficients C_{ij} that determine the initial conditions, the range of ionization disappears as $E \to 0$ for only one of them, namely C_{12}. This means that as $E \to 0$ the ionization region is diminished in such a manner that its dimensionality is reduced by unity. Therefore, it is sufficient to examine the threshold behavior in those cases when only one of the parameters determining the initial conditions is variable and the remainder have fixed values. Such calculations, performed for the ionization of a hydrogen atom by electrons [79-81], confirm both the assumed nature of the drop-off of the ionization region and the threshold dependence (24.31).

Peterkop and Tsukerman [79] have examined the case in which the atomic electron in the initial state moves in a circular orbit with a radius of one atomic unit and the incident electron is confined to the plane of the orbit of the atomic electron. The calculations were performed for values of the total orbital angular momentum L = 0 and L = 1. The variable parameter was the angle ϕ determining the position of the atomic electron in the orbit at the initial time $t = t_0$. The final energy of the incident electron (the energy as $t \to \infty$) was determined by numerical integration of the equations of motion. An example of the results is given in Fig. 6. For values of ϕ such that $0 \le \varepsilon(\phi) \le E$, ionization occurs. Figure 7 shows the dependence of the ionization ranges on E. For clarity, we plot S_ϕ/E. The good agreement with (24.31) is evident. Similar results for $L \ge 0$ were obtained by Banks *et al.* [80] and Grujić [81], who used elliptical orbits to describe the initial state of the atomic electron. Banks *et al.* chose the incident-electron impact parameter as the variable parameter, and Grujić used the angle determining the orientation of the elliptical orbit.

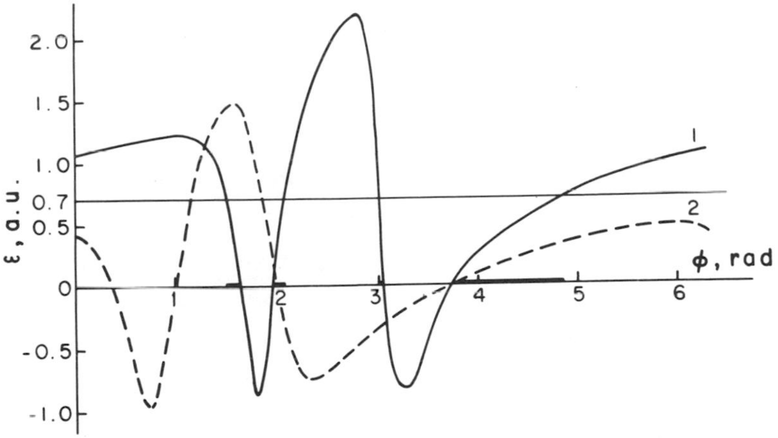

Figure 6. Final energy of the incident electron as a function of the initial polar angle of the atomic electron for L = 1: The solid curve is for E = 0.7 a.u.; the dashed curve is for E = 0.05 a.u. The ionization ranges for E = 0.7 a.u. are indicated by the heavy line.

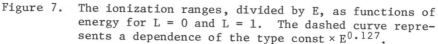

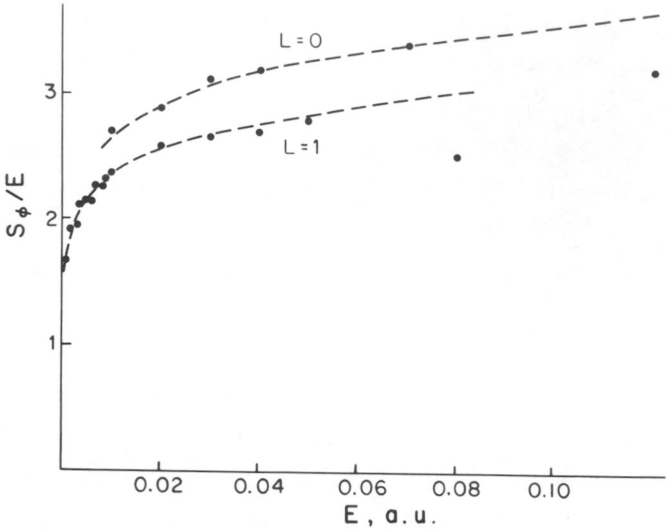

Figure 7. The ionization ranges, divided by E, as functions of energy for L = 0 and L = 1. The dashed curve represents a dependence of the type const $\times E^{0.127}$.

The shape of the curves shown in Fig. 6 changes little as E is varied; however, owing to the nonlinearity of the threshold behavior they should intersect the $\varepsilon = 0$ axis at a right angle for $E = 0$.

The derivative $d\phi/d\varepsilon$ has the meaning of a differential ionization cross section. Since the curve $\varepsilon(\phi)$ is smooth, for $E > 0$ the derivative $d\phi/d\varepsilon$ can be considered constant in this ionization range if E is sufficiently small. Thus, in this case Eq. (24.37) has an obvious interpretation: the curve $\varepsilon(\phi)$ remains smooth for $E \to 0$.

It follows from the existence of a maximum at $\theta = \pi$ that electrons leave the atom with large angular momenta: $l_{max} \sim E^{-1/4}$ [178]. This prediction, as well as the existence of a minimum in the differential cross section at $|E| \approx 0$ which follows from (24.38) and (24.40), was confirmed by classical calculations [179].

§25. *Semiclassical Approximation in the Multidimensional Case*

The semiclassical approximation is applicable to the problem of ionization when the distances between the particles are large, so that the potential varies slowly. In the first place, the semiclassical approximation can be used to determine the asymptotic form of the wave function. It is well known that for scattering by a three-dimensional Coulomb potential, the asymptotic form corresponds to the semiclassical approximation [8,60]. For the semiclassical approximation to be valid the potential must be smooth, although the condition $|V| \ll E$ is not absolutely necessary. This implies that the semiclassical approximation is valid not only in the free-particle domain, but also in a nearer domain -- in the Coulomb zone defined in §24 -- and hence can be used to determine the threshold behavior of the ionization cross section. In this case the advantage of the semiclassical approximation is that its domain of validity does not become infinitely removed as $E \to 0$.

For a system of particles we have a Schrödinger equation in a multidimensional configuration space. The semiclassical approximation in the multidimensional case is developed in analogy with the one-dimensional case. We proceed from the Schrödinger equation, which we write in the form corresponding to (7.25)

$$\left(-\frac{\hbar^2}{2}\Delta + V(\mathbf{r}) - E\right)\Psi(\mathbf{r}) = 0 \quad , \tag{25.1}$$

where $\mathbf{r}$ is an n-dimensional vector and Δ is the n-dimensional Laplacian. We look for a solution of Eq. (25.1) in the form

$$\Psi = \exp\left(\frac{i}{\hbar} S_0 + iS_1 + i\hbar S_2 + \ldots\right) \quad . \tag{25.2}$$

Substituting (25.2) into (25.1) and equating to zero the coefficients of the powers of $\hbar$, we obtain the equations

$$(\nabla S_0)^2 = 2(E - V) \quad , \tag{25.3}$$

$$2(\nabla S_0)\cdot(\nabla S_1) = i\Delta S_0 \quad , \tag{25.4}$$

$$2(\nabla S_0)\cdot(\nabla S_j) = i\Delta S_{j-1} - \sum_{m=1}^{j-1} (\nabla S_m)\cdot(\nabla S_{j-m}) \, , \; j = 2,3,\ldots \quad , \tag{25.5}$$

where ∇ denotes the n-dimensional gradient.

If we introduce the notation

$$R_0 = \exp(iS_1) \quad , \tag{25.6}$$

Eq. (25.4) can be written in the form

$$\nabla\cdot(R_0^2 \nabla S_0) = 0 \quad . \tag{25.7}$$

We obtain the semiclassical approximation by retaining the first two terms of expansion (25.2); i.e., it is defined by Eqs. (25.3) and (25.7). Equation (25.3) is the same as the Hamilton-Jacobi equation, and Eq. (25.7) is the equation of flux conservation in classical mechanics. If S_0 and R_0 are real, then they describe the motion of some classical ensemble of particles. The function S_0 is the classical action and determines the velocity of the classical flux (we set the masses of the particles equal to unity)

$$\underset{\sim}{v}(\underset{\sim}{r}) = \nabla S_0(\underset{\sim}{r}) \quad . \tag{25.8}$$

R_0 determines the particle density

$$P(\underset{\sim}{r}) = [R_0(\underset{\sim}{r})]^2 \quad . \tag{25.9}$$

The classical flux vector is

$$\underset{\sim}{j} = P\underset{\sim}{v} = R_0^{\,2}\nabla S_0 \quad . \tag{25.10}$$

The wave function in the semiclassical approximation takes the form of a linear combination

$$\Psi_{c\ell} = \sum_j c_j R_0^{(j)} \exp(iS_0^{(j)}/\hbar) \quad , \tag{25.11}$$

where $S_0^{(j)}$ is a particular solution of Eq. (25.3) and $R_0^{(j)}$ is the corresponding solution of Eq. (25.7). The coefficients c_j are determined from the boundary conditions and the symmetry properties. If the sum (25.11) contains only one term, then the quantum mechanical expressions for the particle and current densities agree with Eqs. (25.9) and (25.10). In this situation the semiclassical approximation is identical to the classical description. If (25.11) contains several terms, then the quantum mechanical expressions differ from the classical expressions by the presence of interference terms. In general, the classical ensemble also describes all of the various solutions of (25.3) and (25.7), but in this description there is no interference. Another feature of the semiclassical approximation is that motion in classically forbidden regions can be considered. Through the boundary conditions these regions can also affect the wave function in the allowed regions. Here we do not consider the semiclassical approximation in the forbidden regions and the problems of matching the wave function at turning points, which in the multidimensional case form hypersurfaces known as caustics. We remark that the functions that approximate the exact solution of the Schrödinger equation in a finite neighborhood of the caustics were obtained by Khudyakov [82] for the case of three-dimensional potential scattering.

The functions S_0 and R_0 directly represent the velocity and density of an ensemble of particles, but not the motion of an individual particle along a trajectory. In this sense Eqs. (25.3) and (25.7) express the formulation of classical mechanics that is closest to quantum mechanics. However, from S_0 it is not difficult to find also the trajectories of individual particles. They are determined by Eq. (25.8), which can be rewritten in the form

$$\frac{d\underset{\sim}{r}}{dt} = \nabla S_0(\underset{\sim}{r}) \quad , \qquad (25.12)$$

where t is the time. Thus, each function S_0 determines some family of trajectories (transversals of S_0), along which the motion of the given classical ensemble takes place.

The density R_0^2 is determined by three factors. It is proportional to the density of trajectories and inversely proportional to the velocity of the motion; each individual trajectory can also have an additional density factor, which expresses the fact that if R_0^2 is some solution of Eq. (25.7), then the product ΦR_0^2 will also be a solution under the condition that Φ satisfy the equation

$$(\nabla\Phi)\cdot(\nabla S_0) = 0 \quad . \qquad (25.13)$$

Since ∇S_0 determines the direction of the trajectory, Φ is constant along the trajectory.

There is no need to solve (25.7) and (25.12) if S_0 is known in the form of a complete integral of Eq. (25.3)

$$S_0 = S_0(\alpha_1,\ldots,\alpha_n;\ x_1,\ldots,x_n) \quad , \qquad (25.14)$$

where one of the parameters α_j denotes the energy E, and the remaining are nonadditive constants of integration. The set $\alpha_1,\ldots,\alpha_n$ can also signify any other set of linearly independent combinations of these parameters. We obtain the equations of the trajectories by differentiating the function [75]

$$\tilde{S} = S_0 - Et \tag{25.15}$$

with respect to the parameters α_j and equating to new constants

$$\frac{\partial \tilde{S}}{\partial \alpha_j} = \gamma_j \quad . \tag{25.16}$$

A particular solution to the equation of continuity (25.7) is the van Vleck determinant [83]

$$R_0^{\ 2} = \left\| \frac{\partial^2 S_0}{\partial x_i \partial \alpha_j} \right\| \quad . \tag{25.17}$$

If arbitrary coordinates $q_1, \ldots, q_n$ are used, then the right side of (25.17) should be divided by the factor g that enters the expression $d\underset{\sim}{r} = g dq_1 \ldots dq_n$ for the volume element.

In order to clarify the conditions under which the semiclassical approximation is valid, we consider a somewhat different derivation of Eqs. (25.3) and (25.7) [84]. We look for a solution to the Schrödinger equation in the form

$$\Psi = R \exp(iS/\hbar) \quad , \tag{25.18}$$

where R and S are real. Putting (25.18) into (25.1), we obtain equations analogous to (8.12) and (8.13),

$$(\nabla S)^2 = 2\left(E - V + \frac{\hbar^2}{2} \frac{\Delta R}{R}\right) \quad , \tag{25.19}$$

$$\nabla \cdot (R^2 \nabla S) = 0 \quad . \tag{25.20}$$

The reality condition on R and S is not essential. The function (25.18) will also satisfy (25.1) if R and S are complex solutions of (25.19) and (25.20). Equations (25.19) and (25.20) go over to (25.3) and (25.7) if we drop the term containing $\hbar^2$ in (25.19). The condition under which we may make this approximation is

$$\frac{\hbar^2}{2}\left|\frac{\Delta R_0}{R_0}\right| \ll |E - V| \quad . \tag{25.21}$$

This condition is verified by substituting the semiclassical wave function into (25.1). Taking (25.3) and (25.7) into account, we find

$$\left(-\frac{\hbar^2}{2}\Delta + V - E\right)R_0 \exp\frac{iS_0}{\hbar} = -\frac{\hbar^2}{2}(\Delta R_0)\exp\frac{iS_0}{\hbar} \quad . \tag{25.22}$$

Note that the condition

$$\frac{1}{4\pi}\left|\frac{d\lambda}{dx}\right| \ll 1 \quad , \tag{25.23}$$

which is often used in the one-dimensional case (λ is the deBroglie wavelength), is equivalent to

$$\hbar\left|\frac{dS_1}{dx}\right| \ll \left|\frac{dS_0}{dx}\right| \quad . \tag{25.24}$$

The condition (25.21) in the one-dimensional case is equivalent to

$$2\hbar^2\left|\frac{dS_2}{dx}\right| \ll \left|\frac{dS_0}{dx}\right| \quad . \tag{25.25}$$

The question of the validity of the semiclassical approximation in the multidimensional case is more complicated than in the one-dimensional case. The condition (25.21) states that semiclassical mechanics is applicable if the density of the particle ensemble varies sufficiently slowly. In the one-dimensional case the density is uniquely related to the potential, because for $n = 1$ we have $R_0 = \text{const}(E - V)^{-1/4}$ and (25.21) reduces to the condition that the potential changes slowly. In the multidimensional case there is a significantly greater arbitrariness in the choice of solutions, and there is no unique relation between the density and the potential. The slow variation of the potential is still necessary, but it is no longer a sufficient condition for the applicability of the semiclassical approximation. Semiclassical mechanics is more accurate for solutions of the plane-wave type, when the density varies slowly, and is less so for solutions of the diverging-wave type, when the density varies more rapidly. For the second type of solutions semiclassical mechanics even with $V = 0$ is valid only at large distances.

The asymptotic form of the semiclassical wave function agrees with (6.39), (9.1), and (9.2). The classical equations (25.3) and (25.7) follow from the quantum equations (25.19) and (25.20), if we set $\hbar = 0$. Similarly, we obtain the equations for R_0 and S_0 in hyperspherical coordinates by setting $\hbar = 0$ in (8.12) and (8.13). Thus it is clear that the functions R_0 and S_0 can be expanded in series of the form (8.15) and (8.16). We obtain the expansion coefficients by setting $\hbar = 0$ in the coefficients of the quantum case. We note that in the semiclassical case the recursion relations are simplified because the terms in the expansion for S_0 are determined independently from those in the expansion for R_0. Keeping only the leading terms, we find for the asymptotic forms

$$R_0 \sim A(\hat{\Omega}) r^{(n-1)/2} , \tag{25.26}$$

$$S_0 \sim \kappa r + \frac{Z(\hat{\Omega})}{\kappa} \ln \kappa r , \tag{25.27}$$

where A is the scattering amplitude, n is the dimension of the configuration space of the reduced particles in the center-of-mass system, $\kappa = \sqrt{2E}$, and r and Z are determined by Eqs. (7.29) and (6.1).

Another possible way to obtain the asymptotic form is by iteration. Rudge and Seaton [6] show that the asymptotic form (25.27) follows if the centrifugal term in the Hamilton-Jacobi equation is treated iteratively. Equation (25.27) can also be obtained by an iterative treatment of the Coulomb potential energy in the classical equations of motion [38]. Newton's equations for a system of charged particles are of the form

$$m_i \frac{d^2 r_i}{dt^2} = \zeta_i \sum_{j \neq i} \frac{\zeta_j (\underset{\sim}{r}_i - \underset{\sim}{r}_j)}{|\underset{\sim}{r}_i - \underset{\sim}{r}_j|^3} \quad , \tag{25.28}$$

where m_i is the mass and ζ_i the charge of the *ith* particle. In the initial approximation we drop the right side of the equation. This gives

$$\underset{\sim}{r}_i^{(0)} = \underset{\sim}{v}_i t + \underset{\sim}{c}_i \quad , \tag{25.29}$$

where $\underset{\sim}{v}_i$ and $\underset{\sim}{c}_i$ are constants. Equation (25.29) corresponds to free motion. Substituting (25.29) into the right side of (25.28) and integrating, we find

$$\underset{\sim}{r}_i^{(1)} = \underset{\sim}{v}_i t - \frac{\zeta_i}{m_i} \sum_{j \neq i} \frac{\zeta_j (\underset{\sim}{v}_i - \underset{\sim}{v}_j)}{|\underset{\sim}{v}_i - \underset{\sim}{v}_j|^3} \ln t + O(1) \quad . \tag{25.30}$$

Subsequent iterations give corrections which remain finite for $t \to \infty$.

The solution of Eq. (25.3) is the classical or Lagrange action; thus it can be represented in the form

$$S_0 = 2 \int T dt \quad , \tag{25.31}$$

where T is the kinetic energy,

$$T = \frac{1}{2}\sum_j m_j \left(\frac{d\underset{\sim}{r}_j}{dt}\right)^2 \quad . \tag{25.32}$$

Using (25.30) we obtain

$$S_0^{(1)} = \kappa^2 t + 2\,\frac{Z(\hat{\Omega})}{\kappa}\,\ln t + O(1) \quad , \tag{25.33}$$

where

$$\kappa^2 = \sum_i m_i v_i^2 \, , \quad \frac{Z(\hat{\Omega})}{\kappa} = -\sum_{j<i} \frac{\zeta_i \zeta_j}{|\underset{\sim}{v}_i - \underset{\sim}{v}_j|} \quad . \tag{25.34}$$

On the other hand, from (25.30) we find

$$r \equiv \left(\sum_i m_i r_i^2\right)^{1/2} = \kappa t + \frac{Z(\hat{\Omega})}{\kappa^2}\,\ln t + O(1) \quad . \tag{25.35}$$

Comparison of (25.33) and (25.35) gives (25.27).

In §27 it is shown that when the solutions of Eqs. (25.3) and (25.7) are specified in the neighborhood of some direction, the semiclassical approximation verifies the asymptotic expression (6.39) in the form of a series expansion in powers of the deviation from the direction in question.

§26. *Semiclassical Expressions for Multidimensional Coulomb Wave Functions*

Some idea of the validity of the semiclassical approximation for a multidimensional Coulomb interaction is afforded by the multidimensional Coulomb problem with $Z(\hat{\Omega})$ = const, for which, as in the quantum case, Eqs. (25.3) and (25.7) can be solved in analytical form, following Peterkop and Shkele [56].

The Hamilton–Jacobi equation (25.3) is of the form

$$\sum_{j=1}^{n} \left(\frac{\partial S}{\partial x_j}\right)^2 = \kappa^2 + \frac{2Z}{r} \quad . \tag{26.1}$$

The function S(r) below represents the action which in §25 was denoted by S_0. Equation (26.1) has a solution of the form

$$S = \underset{\sim}{k}\cdot\underset{\sim}{r} + y_1(\eta) = \kappa r + y_2(\eta) \;,\; |\underset{\sim}{k}| = \kappa \quad , \tag{26.2}$$

where η is defined by (16.14). Substituting (26.2) into (26.1), we find

$$\left(\frac{dy_1}{d\eta}\right)^2 - \frac{dy_1}{d\eta} - \frac{Z}{\kappa\eta} = 0 \quad . \tag{26.3}$$

This implies

$$\frac{dy_1}{d\eta} = \frac{1}{2}(1 \mp u) \quad , \tag{26.4}$$

where

$$u = \sqrt{1 + 4Z/\kappa\eta} \quad . \tag{26.5}$$

We find the following two solutions which correspond to the two signs in (26.4):

$$S_1 = \underset{\sim}{k}\cdot\underset{\sim}{r} + y \;,\quad S_2 = \kappa r - y \quad , \tag{26.6}$$

where

$$y = \frac{1}{2}\eta(1-u) - \frac{Z}{\kappa}\,\ell n\,\frac{\kappa\eta(1+u)^2}{4|Z|} \quad . \tag{26.7}$$

The constant of integration in (26.7) results from integrating (26.4) between the limits 0 and η, which ensures finite values for $\kappa \to 0$. For $\kappa = 0$ and $Z > 0$, we have

$$S_1 = -\sqrt{8Zr}\,\sin\frac{\theta}{2}\,,\quad S_2 = \sqrt{8Zr}\,\sin\frac{\theta}{2}\quad, \tag{26.8}$$

where θ is defined by (16.14).

In the asymptotic domain ($\eta\to\infty$) S_1 and S_2 are equal, up to a constant, to the phases of the incident and scattered waves of the quantum Coulomb wave function, given by (16.18)-(16.20). S_1 describes the family of trajectories that comprise at infinity a pencil parallel to the direction $\underset{\sim}{k}$ (with logarithmic phase shift). S_2 describes a radially diverging bundle of trajectories.

It is convenient to find the equations for the trajectories from Eq. (25.16) by differentiating with respect to the components of the wave vector $\underset{\sim}{k}$, which gives

$$x_j + \frac{1}{2}(1\mp u)\left(\frac{k_j}{\kappa}\,r - x_j\right) \pm \frac{Zk_j}{\kappa^3}\,\ell n\,\frac{\kappa\eta(1+u)^2}{4|Z|} = k_j t + \gamma_j\,. \tag{26.9}$$

If the x_1 axis is chosen in the direction of $\underset{\sim}{k}$, then

$$k_1 = \kappa\,,\quad k_2 = \ldots = k_n = 0\,,\quad \eta = \kappa(r - x_1) \tag{26.10}$$

and (26.9) for $j \geq 2$ takes the form

$$x_j = \frac{2\gamma_j}{1\pm u}\quad. \tag{26.11}$$

We note that Eqs. (26.6)-(26.11) do not depend on n, the dimension of the space. Equation (26.11) determines the trajectories in coordinate space. The bundle of trajectories is axially symmetric, since the right side of (26.11) depends only on r and θ. Each trajectory lies in a two-dimensional plane, which also contains the vector $\underset{\sim}{k}$. The trajectories are hyperbolas with a general asymptote in the direction opposite to $\underset{\sim}{k}$. The parameter γ_1 determines the origin of the time coordinate.

We shall assume $\gamma_1 = 0$; the remaining γ_j determine the impact parameter. In considering a trajectory for given $\gamma_2,\ldots,\gamma_n$, it is convenient to choose the x_2 axis such that

$$\gamma_2 = \gamma \ , \quad \gamma_3 = \ldots = \gamma_n = 0 \quad . \tag{26.12}$$

Then the trajectory will lie in the x_1x_2 plane, and (26.11) can be expressed in the form

$$\frac{Z}{\kappa^2}\,(r+x_1) + \gamma(x_2-\gamma) = 0 \quad . \tag{26.13}$$

Figure 8 presents examples of families at trajectories for Coulomb attraction ($Z=1$) and repulsion ($Z=-1$). In the repulsive case all trajectories behave as though reflected from the caustic, which is a hypersurface represented by the equation

$$\eta = \frac{4|Z|}{\kappa} \quad , \tag{26.14}$$

on which $u=0$. In the x_1x_2 plane Eq. (26.14) determines a parabola. The region where $\eta < 4|Z|/\kappa$ is the classically excluded (shadow) region. It includes the energetically forbidden region where $r < 2|Z|/\kappa^2$. In the classically allowed region, through every point of space pass two trajectories which belong respectively to S_1 and S_2. The trajectories of the two families merge at the caustic, where

$$\nabla S_1 = \nabla S_2 \quad . \tag{26.15}$$

The two possible signs in (26.9) and (26.11) correspond to the two branches of the same hyperbola. Thus, classical Coulomb scattering is described by two noninterfering fluxes. The existence of only one smooth caustic is a peculiarity of the Coulomb potential. In the general case of an arbitrary potential the shape of the caustic is more complicated; the number of reflected fluxes can increase and, hence, also the number of

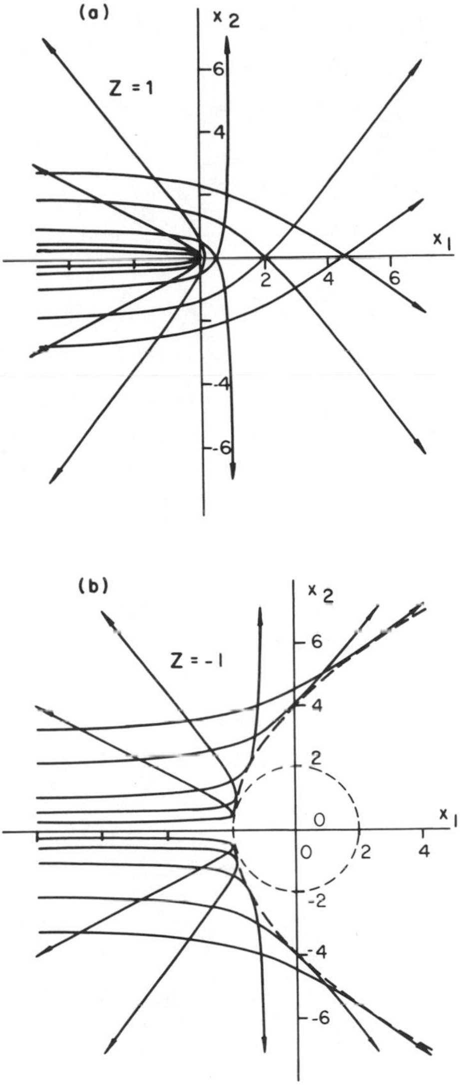

Figure 8. Families of trajectories corresponding to a wave function with the asymptotic form "plane wave + diverging wave."

functions S_j which describe classical scattering. We note that for the constants of integration chosen in (26.6) and (26.7) S_1 and S_2 are also equal on the caustic.

For an attractive potential ($Z > 0$) there is no classically forbidden region. In this case the role of the caustic is assumed by the positive x_1 axis ($\eta = 0$), through which all trajectories pass. From Fig. 8 it is evident that the scattering angles in the cases of Coulomb attraction and repulsion are the same in absolute value, but opposite in direction. The differential scattering cross sections are identical.

The classical functions (26.6) and (26.7) correspond to the phases of the quantum wave function (16.18), which has the asymptotic behavior "plane wave + diverging wave" (the words "plane wave" should be understood as implying a logarithmically phase-shifted plane wave). In order to obtain the classical action functions corresponding to the quantum wave function $\psi^{(-)}$ with the asymptotic behavior "plane wave + converging wave," it is necessary to change the sign of the function S_j and of the direction $\underset{\approx}{k}$ in (26.6) and (26.7). The functions given by these equations will still be solutions of Eq. (26.1). We thus obtain

$$S_1^{(-)} = \underset{\approx}{k}\cdot\underset{\approx}{r} - \tilde{y}\,, \quad S_2^{(-)} = -\kappa r + \tilde{y} \quad , \tag{26.16}$$

where

$$\tilde{y} = \frac{1}{2}\tilde{\eta}(1-\tilde{u}) - \frac{Z}{\kappa}\,\ln\frac{\kappa\tilde{\eta}(1+\tilde{u})^2}{4|Z|} \quad , \tag{26.17}$$

$$\tilde{u} = \sqrt{1+4Z/\kappa\tilde{\eta}}\;, \quad \tilde{\eta} = \kappa r + \underset{\approx}{k}\cdot\underset{\approx}{r} \quad . \tag{26.18}$$

Changing the sign of the action function S_j changes the direction of motion along the trajectory, and changing the direction of $\underset{\approx}{k}$ is equivalent to a reflection in the x_2 axis. As a result we find the family of trajectories shown in Fig. 9. In this case we have sets of trajectories entering from different directions and leaving parallel to a given direction. It seems natural that just such a set should describe the final state for scattering in a given direction.

Let us consider further the flux conservation equation (25.7), which we rewrite in the form

$$\sum_{j=1}^{n} \frac{\partial}{\partial x_j}\left(P \frac{\partial S}{\partial x_j}\right) = 0 \quad , \tag{26.19}$$

where $P = R_0^2$ denotes the particle density. The solution of (26.19) can be found by means of (25.17), but in the present case it is simpler to solve (26.19) directly. Substituting (26.2) into (26.19) and assuming that P depends only on η, we obtain the equation

$$\eta u \frac{dP}{d\eta} + \left[\frac{n-1}{2}(u \mp 1) + \eta \frac{du}{d\eta}\right] P = 0 \quad . \tag{26.20}$$

The case in which P depends only on η corresponds to the quantum case in which the modulus of the wave function (16.13) likewise depends only on η. The solutions of (26.20) take the form

$$P_1 = \frac{1}{u}\left(\frac{1+u}{2}\right)^{n-1} \quad , \tag{26.21}$$

$$P_2 = \frac{1}{u}\left(\frac{|1-u|}{2}\right)^{n-1} \quad . \tag{26.22}$$

Equations (26.6), (26.7), (26.21), and (26.22) for n = 3 agree with the semiclassical solution given by Khudyakov [82]. In contrast with (26.6) and (26.7), Eqs. (26.21) and (26.22) depend on the dimensionality of the space. P_1 and P_2 refer to S_1 and S_2; i.e., P_1 is the particle density in the incident flux and P_2 is the density in the scattered flux. The normalization factor

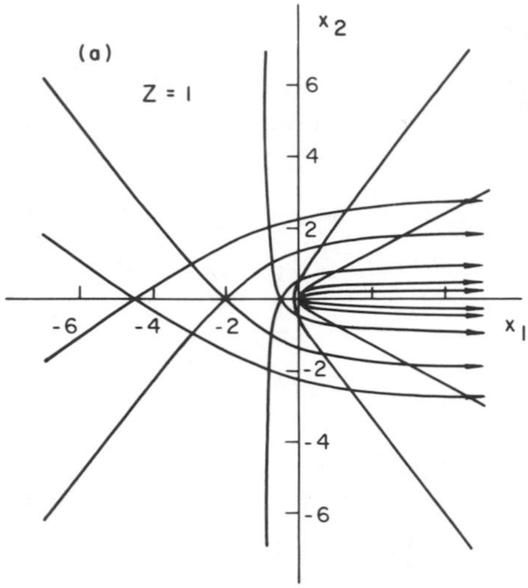

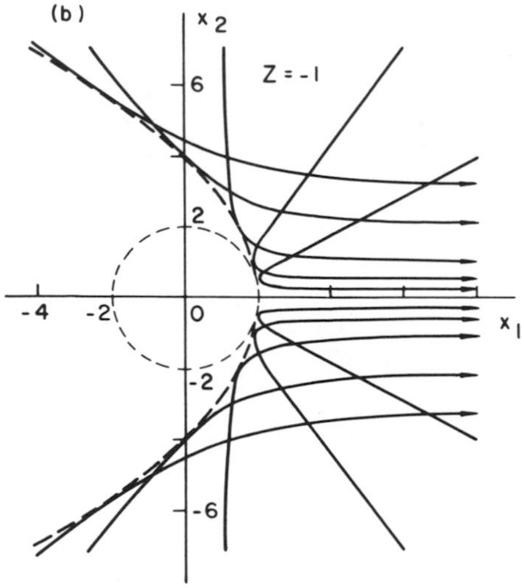

Figure 9. The families of trajectories corresponding to the wave function with asymptotic behavior "plane wave + converging wave."

for P_1 is chosen such that $P_1 \to 1$ as $\eta \to \infty$. The density P_2 is normalized to coincide with P_1 on the caustic. This ensures a continuous transition from one flux to the other. The trajectories become more concentrated on the caustic. The densities P_1 and P_2 go to infinity on the caustic, but their ratio goes to unity.

The classical differential scattering cross section can be defined similarly to the quantum cross section as the ratio of the scattered and incident fluxes,

$$\frac{d\sigma}{d\hat{\Omega}} = \frac{j_2}{j_1}\frac{dS}{d\hat{\Omega}} = \frac{P_2}{P_1} r^{n-1} \quad . \tag{26.23}$$

For $\eta \to \infty$ we obtain

$$\frac{d\sigma}{d\hat{\Omega}} = \lim_{r\to\infty} r^{n-1} P_2 = \left[\frac{|Z|}{2\kappa^2 \sin^2(\theta/2)}\right]^{n-1} \quad , \tag{26.24}$$

which agrees with the quantum cross section (16.24) and (16.25) only for n = 3; for other values of n it differs by a factor independent of θ. We note that this factor goes to unity for $\kappa \to 0$.

In using (26.23), it is not necessary to know the trajectories of individual particles, but the result, of course, must agree with the result obtained by the method using the trajectories, which we also present here. Considering (26.11) for large η, we obtain a relation between the scattering angle and the impact parameter

$$\text{tg}\,\frac{\theta}{2} = \frac{|Z|}{\gamma\kappa^2} \quad . \tag{26.25}$$

Since σ is the volume of an n-1 dimensional subspace, we have

$$d\sigma = \omega_{n-1}\gamma^{n-2} d\gamma \quad , \tag{26.26}$$

where ω_n is defined by (7.10). Equation (7.9) then gives

$$d\hat{\Omega} = \omega_{n-1}(\sin\theta)^{n-2}d\theta \quad . \tag{26.27}$$

Hence it follows that

$$\frac{d\sigma}{d\hat{\Omega}} = \left(\frac{\gamma}{\sin\theta}\right)^{n-2}\left|\frac{d\gamma}{d\theta}\right| = \left[\frac{|Z|}{2\kappa^2\sin^2(\theta/2)}\right]^{n-1} \quad , \tag{26.28}$$

which agrees with (26.24).

According to (25.21) the validity of the semiclassical approximation is determined by the behavior of the density of particles in the classical ensemble. This approximation is not valid in the neighborhood of the caustic, where the density goes to infinity. The density of particles varies slowly at large distances from the caustic, i.e., for large η. From (26.21) and (26.22) we obtain

$$\frac{\Delta\sqrt{P_1}}{\sqrt{P_1}} = -\frac{(n-3)(n-5)Z}{2r\eta^2} + \frac{(2n^2 - 19n + 43)Z^2}{\kappa r\eta^3} + \ldots \quad , \tag{26.29}$$

$$\frac{\Delta\sqrt{P_2}}{\sqrt{P_2}} = \frac{(n-1)\kappa}{r\eta} - \frac{(n^2 + 6n + 1)Z}{2r\eta^2} + \ldots \quad . \tag{26.30}$$

We see that for the incident wave the domain of validity sets in earlier than for the scattered wave, especially in the cases of $n = 3$ and $n = 5$. This is associated with the difference, mentioned earlier, in the validity of the semiclassical approximation for solutions behaving approximately as plane waves and those behaving as spherically diverging waves.

The actual expansion parameter in (26.29) and (26.30) is $Z/\kappa\eta$, so these expansions become invalid as $\kappa \to 0$. For Coulomb repulsion ($Z < 0$) the classically forbidden region as $\kappa \to 0$ extends over the whole space ($\eta_{min} = 4|Z|/\kappa \to \infty$). This is associated with the exponential drop-off of the quantum-mechanical coefficient C_n, defined by Eqs. (16.17), (16.28), and (16.29). For the attractive case ($Z > 0$) the wave function for any given r goes to infinity for $\kappa \to 0$; the functions S_1 and S_2 are given by Eq. (26.8), and for the density we find

$$P_1 \approx P_2 \approx \left(\kappa \sin \frac{\theta}{2}\right)^{2-n} \left(\frac{2Z}{r}\right)^{n/2-1} \quad , \quad Z > 0 \quad . \tag{26.31}$$

By comparing Eq. (26.31) with Eq. (16.30) it follows that P_1, P_2 and $|C_n|^2$ all depend on κ to the same power. From a comparison of (26.8) and (26.31) with (16.30) it follows that, as $\kappa \to 0$, the semiclassical approximation agrees with the asymptotic form of the quantum wave function for large r. Thus, for $\kappa = 0$ and $Z > 0$ the semiclassical approximation is valid for sufficiently large r. From (26.31) we obtain

$$\frac{\Delta\sqrt{P_j}}{\sqrt{P_j}} \approx \frac{n-2}{16r^2}\left[4 - n - (n-2)\operatorname{ctg}^2 \frac{\theta}{2}\right] \quad . \tag{26.32}$$

The ratio of (26.32) to the Coulomb potential energy falls off as r^{-1}. Hence, for $Z(\hat{\Omega})$ = const the error in the semiclassical approximation for $\kappa = 0$ decreases as r^{-1} for increasing r.

The classical trajectories for $\kappa = 0$ and $Z > 0$ are determined by the equation

$$\sqrt{r} \cos \frac{\theta}{2} = \text{const} \quad . \tag{26.33}$$

In this case the trajectories and lines of constant action in the x_1x_2 plane are mutually perpendicular parabolas, which are symmetric with respect to the x_1 axis. Owing to the symmetry of the two branches of trajectories, the incident and scattered fluxes are also mutually symmetric. Their trajectories are the same, but the directions of motion are different.

§27. *Semiclassical Approximation in Low-Energy Ionization*

The semiclassical approximation has been used to determine the threshold behavior of the ionization cross section in the cases in which the motion of the electrons is either one dimensional [85-87] or three dimensional, with total orbital angular momentum L = 0. In essence, the semiclassical treatment is equivalent to the classical treatment of Wannier but has the advantage that it enables us to use the more familiar methods of quantum mechanics.

The Hamilton-Jacobi equation (25.3) and the flux-conservation equation (25.7) for two electrons in the field of a fixed nucleus are

$$(\nabla_1 S)^2 + (\nabla_2 S)^2 = 2E + \frac{2Z(\alpha,\theta)}{r} \quad , \qquad (27.1)$$

$$\nabla_1 \cdot (P\nabla_1 S) + \nabla_2 \cdot (P\nabla_2 S) = 0 \quad , \qquad (27.2)$$

where ∇_1 and ∇_2 are three-dimensional operators acting on $\underset{\sim}{r}_1$ and $\underset{\sim}{r}_2$, and $Z(\alpha,\theta)$ is defined by Eq. (24.1).

We shall consider the case in which L = 0; then S and P depend only on r, α, θ. Equations (27.1) and (27.2) take the form

$$\left(\frac{\partial S}{\partial r}\right)^2 + \frac{1}{r^2}\left(\frac{\partial S}{\partial \alpha}\right)^2 + \frac{4}{r^2 \sin^2 2\alpha}\left(\frac{\partial S}{\partial \theta}\right)^2 = 2E + \frac{2Z(\alpha,\theta)}{r} \quad , \qquad (27.3)$$

$$\frac{1}{r^5} \cdot \frac{\partial}{\partial r}\left(r^5 P \frac{\partial S}{\partial r}\right) + \frac{1}{r^2}\left[D_\alpha\left(P \frac{\partial S}{\partial \alpha}\right) + D_\theta\left(P \frac{\partial S}{\partial \theta}\right)\right] = 0 \quad , \qquad (27.4)$$

where

$$D_\alpha f = \frac{1}{\sin^2 2\alpha} \cdot \frac{\partial}{\partial \alpha}\left(f \sin^2 2\alpha\right) \quad , \qquad (27.5)$$

$$D_\theta f = \frac{4}{\sin^2 2\alpha \sin\theta} \cdot \frac{\partial}{\partial\theta} (f \sin\theta) \quad . \tag{27.6}$$

Taking into consideration the properties of the ionization process noted by Wannier (see §24), we look for a solution in the neighborhood of the stationary (saddle) points of the function $Z(\alpha,\theta)$, i.e., at $\alpha \approx \pi/4$ and $\theta \approx \pi$.

We look for a solution to (27.3) in the same form as the expansion (24.6)

$$S = S_0(r) + \frac{1}{2} S_1(r)(\Delta\alpha)^2 + \frac{1}{8} S_2(r)(\Delta\theta)^2 + \ldots \quad . \tag{27.7}$$

Substituting (24.6) and (27.7) into (27.3) we get the equations

$$\frac{dS_0}{dr} = w \quad , \tag{27.8}$$

$$w \frac{dS_i}{dr} + \frac{S_i^2}{r^2} = \frac{Z_i}{r} \quad , \quad i = 1,2 \quad , \tag{27.9}$$

where w is defined by (24.13). The solutions are of the form

$$S_0 = rw + \frac{Z_0}{\kappa} \ell n \frac{r(\kappa+w)^2}{2Z_0} \quad , \quad \kappa = \sqrt{2E} \quad , \tag{27.10}$$

$$S_i = r^2 w \frac{1}{u_i} \cdot \frac{du_i}{dr} \quad , \quad i = 1,2 \quad , \tag{27.11}$$

where u_i is determined by Eqs. (24.15)-(24.18). The constant of integration for S_0 is chosen such that at $E = 0$,

$$S_0 = \sqrt{8Z_0 r} \quad . \tag{27.12}$$

We note that allowing for higher terms in the expansion (27.7) does not change S_0, S_1, and S_2.

The physical meaning of the function S is obtained by considering the family of trajectories which it describes. We show that the function S, taking into account the first three terms of expansion (27.7), corresponds to the trajectories found in §24. According to (25.12) the equations of the trajectories are

$$\frac{d\underset{\sim}{r}_i}{dt} = \nabla_i S \ , \qquad i = 1,2 \quad , \tag{27.13}$$

where t is the time. In terms of the coordinates r, α, θ these equations take the form

$$\frac{dr}{dt} = \frac{\partial S}{\partial r} \ , \quad \frac{d\alpha}{dt} = \frac{1}{r^2} \cdot \frac{\partial S}{\partial \alpha} \ , \quad \frac{d\theta}{dt} = \frac{4}{r^2 \sin^2 2\alpha} \cdot \frac{\partial S}{\partial \theta} \quad . \tag{27.14}$$

Substituting Eqs. (27.7), (27.8), and (27.11) into (27.14), keeping only the first terms of the resulting series, and eliminating t, we find

$$\frac{d\alpha}{dr} = \frac{\Delta\alpha}{u_1} \cdot \frac{du_1}{dr} \ , \quad \frac{d\theta}{dr} = \frac{\Delta\theta}{u_2} \cdot \frac{du_2}{dr} \quad , \tag{27.15}$$

from which Eq. (24.9) follows.

Since S_1 and S_2 contain the functions u_1 and u_2 in the form of logarithmic derivatives, S_1 and S_2 each have only one arbitrary constant; these may be chosen as the ratios

$$c_1 = C_{11}/C_{12} \ , \quad c_2 = C_{21}/C_{22} \quad . \tag{27.16}$$

Equations (24.9) can also be obtained from equations in the form of (25.16):

$$\frac{\partial S}{\partial c_1} = \text{const} \ , \quad \frac{\partial S}{\partial c_2} = \text{const} \quad . \tag{27.17}$$

The derivatives can be calculated by means of the Wronskian

$$u_{i1}\frac{du_{i2}}{dr} - u_{i2}\frac{du_{i1}}{dr} = \frac{\mu_i\sqrt{2Z_0}}{r^2 w} \quad . \tag{27.18}$$

Let us elucidate the physical meaning of c_1 and c_2. For this we rewrite (24.9) in the form

$$\Delta\alpha = C_{12}(c_1 u_{11} + u_{12}) \,, \quad \Delta\theta = C_{22}(c_2 u_{21} + u_{22}) \quad . \tag{27.19}$$

For a given value of c_1, by allowing C_{12} to vary, we see that the first equation of (27.19) describes the family of trajectories which, as functions of r and α, have a common point of intersection for all trajectories. At the point of intersection $\Delta\alpha = 0$. We denote the value of r at this point by R_1. The value of r at the analogous point of intersection of the trajectories, described by the second equation of (27.19), we denote by R_2. The values R_1 and R_2 are determined by the equation

$$c_i u_{i1}(R_i) + u_{i2}(R_i) = 0 \quad . \tag{27.20}$$

Thus, c_i determines the family of trajectories with a definite point of intersection. Note that in general R_i can be complex.

At the intersection point where $r = R_i$ the function S_i has a pole. In the neighborhood of R_i,

$$S_i \approx \frac{r^2 w}{r - R_i} \quad . \tag{27.21}$$

A singularity in S_i does not mean a singularity of the function S; it simply means that at these points the expansion (27.7) is inapplicable. To clarify this we compare (27.21) with the case in which $Z(\alpha,\theta) = 0$. Equation (27.3) then has the exact solution

$$S^{(0)} = \kappa\left[r^2 - rR(2 - 2\cos\theta\sin 2\alpha)^{1/2} + R^2\right]^{1/2} + \kappa R \quad , \tag{27.22}$$

where R is an arbitrary parameter (the coordinate of the point where the trajectories intersect). The series expansion of $S^{(0)}$ is

$$S^{(0)} = \kappa r + \frac{\kappa r R}{r-R}\left[\frac{(\Delta\alpha)^2}{2} + \frac{(\Delta\theta)^2}{8}\right] + \ldots \quad . \qquad (27.23)$$

The singularity in (27.23) is the same as in (27.21) for $Z_0 = 0$. In deriving Eq. (27.23) the root of the expression in brackets in (27.22) is taken with the negative sign for $r < R$ and with the positive sign for $r > R$. This ensures that the direction of the trajectories is preserved at the point of intersection.

The asymptotic behavior of the function S is different when R_1 and R_2 -- the points of intersection of the trajectories -- are at a finite distance or at infinity. The latter occurs when $u_i(\infty) = 0$, $i = 1,2$, which together with (24.22) means

$$c_i\gamma_{i1} + \gamma_{i2} = 0 \ , \ i = 1,2 \quad . \qquad (27.24)$$

If the intersection points of the trajectories are at finite distances, i.e., $u_i(\infty) \neq 0$, then the Eqs. (27.19) describe a spherically diverging pencil of trajectories. In this case for the function S we find, using (24.22), that

$$S \sim \kappa r + \frac{1}{\kappa}\left[Z_0 + \frac{1}{2}Z_1(\Delta\alpha)^2 + \frac{1}{8}Z_2(\Delta\theta)^2\right]\ell n\,\frac{Er}{Z_0} + \ldots \ , \qquad (27.25)$$

which corresponds to a spherically diverging wave. Considering (24.6), we see that the logarithmic term in this expression corresponds to the phase shift in (6.39). Thus, (27.25) confirms the asymptotic form obtained earlier.

If the points of intersection lie at infinity, then from (24.22), (24.25), and (27.24) we obtain

$$c_i u_{i1} + u_{i2} \sim -\frac{\mu_i}{\gamma_{i1}} \sqrt{Z_0/E} \cdot \frac{1}{r} \quad , \tag{27.26}$$

$$S \sim \kappa r \left[1 - \frac{1}{2}(\Delta\alpha)^2 - \frac{1}{8}(\Delta\theta)^2 + \ldots\right] \quad . \tag{27.27}$$

In this case $\Delta\alpha$ and $\Delta\theta \to 0$ as $r \to \infty$; i.e., all trajectories have the same direction ($\alpha = \pi/4$, $\theta = \pi$), so the solution is a plane wave. In contrast with (27.25), here we have found only the first term of the asymptotic form, which does not contain a logarithmic part. The resulting expression should agree with the phase of the plane wave, or more precisely, with the phase of that part of the plane wave that corresponds to $L = 0$. In order to show that this actually does occur, we note that (27.27) is equal to the limit as $R \to \infty$ of Eq. (27.23), which corresponds to free motion. The exact solution [with $Z(\alpha,\theta) = 0$] of Eq. (27.3) in the form (27.22) as $R \to \infty$ is

$$S^{(0)} = \kappa r \sqrt{\frac{1}{2}(1 - \cos\theta \sin 2\alpha)} = \frac{\kappa}{\sqrt{2}} |\underset{\sim}{r}_1 - \underset{\sim}{r}_2| \quad . \tag{27.28}$$

The present case ($\alpha = \pi/4$, $\theta = \pi$ or $\underset{\sim}{k}_1 = -\underset{\sim}{k}_2$) is represented by the plane wave

$$\psi^{(0)} = \exp[i\underset{\sim}{k}_1 \cdot (\underset{\sim}{r}_1 - \underset{\sim}{r}_2)] \ , \quad k_1 = \kappa/\sqrt{2} \quad . \tag{27.29}$$

The eigenfunctions of the operator $\hat{L}^2$ corresponding to the eigenvalue $L = 0$ are the Legendre polynomials $P_\ell(\cos\theta)$, $\ell = 0, 1, \ldots$. The $L = 0$ part of the function $\psi^{(0)}$ is

$$\psi_0^{(0)} = \sum_{\ell=0}^{\infty} \frac{2\ell+1}{16\pi^2} P_\ell(\cos\theta) \int \psi^{(0)} P_\ell(\cos\theta)\, d\hat{\Omega}_1 d\hat{\Omega}_2 \quad . \tag{27.30}$$

We integrate using the expansion of the plane wave in Legendre polynomials and sum the result using the addition theorem for

Bessel functions [26]; the result is

$$\psi_0^{(0)} = \frac{\sin(k_1|\underset{\sim}{r}_1-\underset{\sim}{r}_2|)}{k_1|\underset{\sim}{r}_1-\underset{\sim}{r}_2|} \quad . \tag{27.31}$$

The phase of this expression agrees with (27.28), and the first terms of the expansion of the phase agree with (27.27).

We turn now to the equation of flux conservation (27.4). Its solution can also be represented in series form

$$P = P_0(r) + \text{terms containing } (\Delta\alpha)^2,\ (\Delta\theta)^2 + \ldots \quad . \tag{27.32}$$

We confine ourselves to determining P_0. Substitution of (27.7), (27.8), and (27.11) into (27.4) leads to

$$P_0 = \frac{C}{r^5 w u_1 u_2^2} \quad , \tag{27.33}$$

where C is an arbitrary constant. Equation (27.33) can also be obtained from (25.17), using (27.18) and multiplying the result by a function satisfying (25.13).

The asymptotic behavior of P_0, as in the previous case, is different for a spherically diverging wave and for a plane wave. In the first case we normalize P_0 by the condition

$$P_0 \xrightarrow[r\to\infty]{} \frac{1}{r^5} \quad . \tag{27.34}$$

By means of (24.22) we then obtain

$$P_0 = \frac{\kappa(C_{11}\gamma_{11} + C_{12}\gamma_{12})(C_{21}\gamma_{21} + C_{22}\gamma_{22})^2}{r^5 w(C_{11}u_{11} + C_{12}u_{12})(C_{21}u_{21} + C_{22}u_{22})^2} \quad . \tag{27.35}$$

In the second case, taking (27.31) into account, we require that

$$P_0^{(\infty)} \to \frac{1}{\kappa^2 r^2} \quad . \tag{27.36}$$

Using (27.26) we obtain

$$P_0^{(\infty)} = \frac{\mu_1\mu_2^2 Z_0^{3/2}}{\sqrt{2}\,E^2 r^5 w(\gamma_{12}u_{11} - \gamma_{11}u_{12})(\gamma_{22}u_{21} - \gamma_{21}u_{22})^2} \ . \quad (27.37)$$

In determining the threshold behavior of the ionization cross section, the behavior of S and P_0 as $E\to 0$ is of interest. For E = 0 we have

$$u_{ij} = r^{m_{ij}} \ . \quad (27.38)$$

For $E = 0$, S_0 is determined by (27.12), and S_1 and S_2 are given by

$$S_i = \sqrt{2Z_0 r}\,\frac{m_{i1}c_i + m_{i2}r^{\mu_i}}{c_i + r^{\mu_i}} \ , \quad (27.39)$$

where m_{ij} and μ_i are defined in §24. As is usual for Coulomb attraction, the function P_0 goes to infinity as $E\to 0$. For the potential (24.23) we obtain

$$P_0 \sim E^{1-m_{12}} \ , \quad (27.40)$$

$$P_0^{(\infty)} \sim E^{-5/2+m_{12}} \ . \quad (27.41)$$

We consider the cases when z = 1 and z = 2, and hence $\mathrm{Re}\, m_{2j} = -1/4$.

We shall determine the threshold behavior of the ionization amplitude by two methods: 1) matching the exact solution with the semiclassical solution, and 2) taking the limit under the integral sign in (17.1), where we use the semiclassical wave function for Φ. In the first method we must use the wave function with the asymptotic form of a spherically diverging wave, and in the second method an asymptotic plane wave is required.

The essence of the matching method is the requirement that the function Ψ and its derivatives remain finite for $E \to 0$. We represent the wave function for $r \geq r_0$ (r_0 is the inner edge of the Coulomb zone) as a linear combination of semiclassical functions

$$\Psi = \sum_j a^{(j)} \sqrt{P_0^{(j)}} \exp[iS^{(j)}/\hbar] \quad , \tag{27.42}$$

where the summation is carried out over wave functions having intersection points at finite distances. Considering the asymptotic forms (27.25) and (27.34), we obtain for the ionization amplitude

$$A_{00} = \sum_j a^{(j)} \quad . \tag{27.43}$$

The condition that Ψ be finite as $E \to 0$ together with Eq. (27.40) implies

$$A_{00} \sim E^{(m_{12}-1)/2} \quad . \tag{27.44}$$

In applying the second method -- taking the limit $E \to 0$ in the integral expression -- we note first of all that owing to the conservation of the total orbital angular momentum, the integral expressions (17.8) and (17.10) remain valid if Ψ, Φ, and A_{00} are replaced by their components corresponding to a definite L. The functions considered in this section all pertain to the case $L = 0$.

The semiclassical wave function which behaves asymptotically as a plane wave for $L = 0$ should be constructed so that its asymptotic form for small $\Delta\alpha$ and $\Delta\theta$ agrees with (27.31). Such a function is

$$\Phi = \frac{1}{2i} \sqrt{P_0^{(\infty)}} \left[\exp\left(\frac{i}{\hbar} S^{(\infty)}\right) - \exp\left(-\frac{i}{\hbar} S^{(\infty)}\right)\right] \quad . \tag{27.45}$$

Substitution of (27.45) into (17.8) with allowance for (27.41) then leads to (27.44).

For the differential ionization cross section we obtain

$$\frac{d\sigma}{d\hat{\Omega}} = \frac{\kappa}{k_0} |A_{00}|^2 \sim E^{m_{12}^{-1/2}} \quad . \tag{27.46}$$

The total ionization cross section is

$$\sigma = 2\pi^2 \int \frac{d\sigma}{d\hat{\Omega}} \sin^2 2\alpha \sin\theta \, d\alpha d\theta \quad . \tag{27.47}$$

As Eqs. (27.44) and (27.46) are valid for $\alpha = \pi/4$ and $\theta = \pi$, it is not possible to substitute (27.46) into (27.47).

It was shown in §24 that the differential cross section has a maximum owing to the clustering of trajectories for $\theta \approx \pi$. In the semiclassical formalism we can demonstrate the narrowing of the distribution over θ by making use of the possibility, mentioned in §25, of multiplying P by a function which is constant along the trajectories. If this function is chosen so that at $r = r_0$ it is nonzero only in some interval $\Delta\theta$, then as $r \to \infty$ the length of this interval falls off as $E^{1/4}$ as E decreases. The clustering of the trajectories increases the density, thus leading to the term -1/2 in the exponent of Eq. (27.46). The reason for the appearance of this term is the presence of u_2^2 in the denominator of Eq. (27.33), because $|u_2(\infty)| \sim E^{1/4}$. Thus, the differential cross section as a function of θ has a maximum $\sim E^{-1/2}$ of width $\sim E^{1/4}$. Since the element of angle contains $\sin\theta$, the overall distribution over θ remains finite as $E \to 0$. This is to be expected, because the bunching together of trajectories cannot change the total cross section. The semiclassical solutions thus obtained do not determine the dependence of the differential cross section on α. Numerical calculations (see §24) show that as $E \to 0$ all finite electron energies are equally probable, hence the differential cross section will depend on α as in (24.39). Thus, we may express approximately the differential cross section near the threshold energy in the form

$$\frac{d\sigma}{d\hat{\Omega}} = \frac{E^{m_{12}-1/2} F(E^{-1/4}\Delta\theta)}{\sin 2\alpha} , \quad |\Delta\theta| \le \Delta\theta_{max} , \tag{27.48}$$

$$\frac{d\sigma}{d\hat{\Omega}} = 0 , \quad |\Delta\theta| > \Delta\theta_{max} , \tag{27.49}$$

where

$$\Delta\theta_{max} = \text{const} \cdot E^{1/4} , \tag{27.50}$$

and F is a bounded function ($0 \le F(x) \le \text{const}$).

We substitute Eqs. (27.48)-(27.50) into (27.47). Integrating over θ (using $\sin\theta \approx -\Delta\theta$), we obtain Eq. (24.39), whence (24.28) follows. Thus, semiclassical methods lead to Wannier's result.

If we expand the ionization amplitude in the series (13.30) of K harmonics and consider only the first ($K = 0$) harmonic, we find

$$\frac{d\sigma}{d\hat{\Omega}} = \text{const} , \tag{27.51}$$

where the constant depends on E. The substantial difference in the α- and θ-dependence between Eqs. (27.48) and (27.51) indicates that the K harmonics are poorly suited for handling the specifics of the threshold behavior of the ionization cross section.

In our approximation of using the first three terms in the expansions (24.6) and (27.7), the dependence of the classical trajectories and the semiclassical wave functions on each of the angles is expressed separately. The energy dependence of the total ionization cross section is determined only by the dependence of $Z(\alpha,\theta)$ on α. The θ dependence affects the absolute value of the total cross section and the domain of validity of the threshold law. The classical trajectories have a comparatively smooth character with respect to α and an oscillatory character with respect to θ. Therefore, the application of classical mechanics is more readily justified in the first case. However, even in the oscillatory case, for comparatively large quantum numbers ($n \geq 3$), the classical probability distribution is similar to the mean quantum distribution. Quantization can lead to a discrete set of functions that determine the θ-dependence, but it should not change the threshold dependence of the differential cross section. We note that the classical density goes to infinity at the points where the trajectories intersect, which in the presence of oscillations can be quite numerous. This follows from the fact that in our approximation the oscillation period does not depend on the amplitude.

Wave functions equal to our semiclassical functions have also been obtained by Rau [89]. In contrast with the present exposition Rau proceeds from the Schrödinger equation but expresses the phase of the wave function in the form of the classical action; then by further simplification he obtains solutions that coincide with the semiclassical solutions. Rau did not obtain the correct normalization of the wave function used in the integral representation of the amplitude, and he did not discuss the singularity of the differential cross section. By means of certain supplementary assumptions he obtained the energy dependence of the differential cross section without the $-1/2$ term in the exponent, and the result was carried over to the total cross section.

Roth [90] has applied an approximation, equivalent to the semiclassical approximation, to the Schrödinger equation in a form appropriate to total orbital angular momentum L = 1. It was shown that Wannier's result follows in this case, too.

We note that approximate solutions in the form (27.10), (27.11), and (27.33) satisfy the similarity principle, which for solutions of the Hamilton-Jacobi and flux-conservation equations can be formulated as follows: If $S(1,r,\hat{\Omega})$ and $P(1,r,\hat{\Omega})$ are the solutions for energy E = 1, then the solutions for arbitrary energy E > 0 can be obtained from the equations

$$S(E,r,\hat{\Omega}) = \frac{1}{\sqrt{E}} S(1,Er,\hat{\Omega}) \quad , \qquad (27.52)$$

$$P(E,r,\hat{\Omega}) = p(E)P(1,Er,\hat{\Omega}) \quad , \qquad (27.53)$$

where the function p(E) is determined by the form of boundary conditions.

Let us consider further the asymptotic form of the wave function for E = 0. In this case, as Rudge and Seaton [39] have shown, the Hamilton-Jacobi equation admits a solution in the form

$$S = a(\hat{\Omega})\sqrt{r} \quad . \qquad (27.54)$$

From (27.39) it is seen that this corresponds to the cases in which c_1 and c_2 are equal to zero or infinity. If $c_1 = c_2 = 0$, then

$$a(\hat{\Omega}) = \sqrt{8Z_0}\left[1 + \frac{m_{12}}{4}(\Delta\alpha)^2 + \frac{m_{22}}{16}(\Delta\theta)^2 + \dots\right] \quad . \qquad (27.55)$$

If $c_1 = c_2 = \infty$, then m_{i2} is replaced by m_{i1}. The use of m_{21}, which contains a negative imaginary part, leads to an exponentially growing wave function $\Psi = P^{1/2}\exp(iS/\hbar)$, if $\Delta\theta \neq 0$; m_{11}

leads to plane-wave asymptotic behavior. Therefore, the case $c_1 = c_2 = \infty$ cannot describe the asymptotic behavior of a scattered wave. We note that if $\Delta\theta \neq 0$ the presence of a positive imaginary part in m_{22} causes exponential damping of the wave function which corresponds to the clustering of the classical trajectories at $\theta = \pi$.

From (27.33) for $C_{11} = C_{21} = 0$ and $E = 0$ we obtain

$$P_0 \sim r^{-9/2-m} \quad , \tag{27.56}$$

where

$$m = m_{12} + 2m_{22} \quad . \tag{27.57}$$

Setting $m = 0$ ensures the conservation of flux in the radial direction [39]. However, in contrast to the case of $E > 0$, when $E = 0$ for arbitrarily large distances r the trajectories continue to move away from $\alpha = \pi/4$ and approach $\theta = \pi$; hence, the conservation of the radial flux requires the presence of sources and sinks, which would mean a singularity in the wave function.

The Schrödinger equation (6.3) for $E = 0$ admits a series solution in powers of $\sqrt{r}$ but with no powers of $\ell n\, r$:

$$\Psi = r^{-9/4-m/2} \exp\left(\frac{i}{\hbar}\, a(\hat{\Omega})\sqrt{r}\right) \sum_{n\geq 0} b_n(\hat{\Omega})\, r^{-n/2} \quad . \tag{27.58}$$

Here $a(\hat{\Omega})$ satisfies the equation resulting from the substitution of (27.54) into (27.3),

$$\frac{1}{4} a^2 + \left(\frac{\partial a}{\partial \alpha}\right)^2 + \frac{4}{\sin^2 2\alpha}\left(\frac{\partial a}{\partial \theta}\right)^2 = 2Z(\alpha,\theta) \quad , \tag{27.59}$$

and b_n is determined by the recurrence relation

$$- \frac{m+n}{2} ab_n + \frac{1}{b_n} \left[D_\alpha \left(b_n{}^2 \frac{\partial a}{\partial \alpha} \right) + D_\theta \left(b_n{}^2 \frac{\partial a}{\partial \theta} \right) \right] =$$

$$= i\hbar \left[\frac{1}{4} \left(m + n - \frac{1}{2} \right)^2 - 4 + \Delta^* \right] b_{n-1} \quad , \qquad (27.60)$$

where the operators D_α, D_θ, and Δ^* are defined by Eqs. (27.5), (27.6), and (6.4).

Substituting (27.55) into (27.60) for n = 0 and requiring that $b_0(\hat{\Omega})$ be finite for $\Delta\alpha = \Delta\theta = 0$, we find the condition (27.57).

We obtain the semiclassical approximation by setting $\hbar = 0$ in (27.60). Since $b_{-1} = 0$, b_0 is the same in the quantum and semiclassical cases. Thus, the semiclassical approximation is valid for sufficiently large r when the series (27.58) can be replaced by its first term. If we apply the criterion (25.21) to Eq. (27.33), then we also find that the error in the semiclassical approximation with E = 0 diminishes as r^{-1}.

Methods of determining the threshold behavior of the ionization cross section, which differ from those considered here and in §18, have been proposed by Temkin and co-workers [91,92]. In [91] the threshold behavior of the hydrogen-atom ionization cross section was found by extrapolating certain properties of the doubly-excited states of the H^- ion. The results turned out to be somewhat indefinite. It was found that $\sigma \sim E^m$, where $1.25 \leq m \leq 1.5$. In [92] a simplified model was considered, in which the potential energy of the system was replaced by the expression, valid for $\theta = \pi$,

$$V = - \frac{1}{r_1} - \frac{1}{r_2} + \frac{1}{r_1 + r_2} \quad . \qquad (27.61)$$

The threshold behavior of excitation cross sections for states with large principal quantum number was examined for cases that differed in the use of various approximate expressions for the optical potential. The results were extrapolated to the continuous spectrum. It turns out that different approximations for the optical potential lead to quite different threshold ionization laws. For the exponent m the values 1, 2, and 1/4 were obtained along with dependences of logarithmic and oscillatory types:

$$\sigma \sim E^{5/2}(\ln E)^2 , \quad \sigma \sim E[1 - C\sin(A \ln E + B)] , \qquad (27.62)$$

where A, B, and C are constants.

We note that the potential (27.61) corresponds to a function $Z(\alpha)$ of the form (14.1). If we apply the method of Wannier to this potential, we find for the coefficient Z_2 the value zero and for Z_0 and Z_1 the values given by (24.8) with $z = 1$. The parameters m_{11} and m_{12} retain their previous values; $m_{21} = -1/2$ and $m_{22} = 0$. The differential cross section in this case does not have a maximum at $\theta = \pi$. For any θ we have $d\sigma/d\hat{\Omega} \sim E^{m_{12}}$ instead of Eqs. (27.48)-(27.50). Integrating over α and θ, we obtain the same dependence for the total cross section.

§28. *Experimental Results on the Threshold Behavior of the Ionization Cross Section*

The threshold behavior of the total ionization cross section has been investigated experimentally for helium [93-95], argon [95], hydrogen [96], and the alkali metals [97].

Brion and Thomas [93] compare the experimentally measured ionization cross section for helium with the Wannier equation (24.31) in the range $0 < E \leq 12$ eV. In the interval 6-12 eV both curves are practically straight and with the chosen normalization agree with one another. In the interval 0-5 eV the experimental curve lies above the theoretical curve. Brion and Thomas concluded that the exponent in the threshold law is greater than 1.127. Krige *et al.* [94] repeated the measurements by another method and ascertained that without doubt in the interval 0-2 eV the dependence of the cross section on E is nonlinear and that within the experimental error it agrees with the Wannier equation. Marchand *et al.* [95] examined the dependence of the exponent m on the size of the interval, in which the experimental helium ionization cross section is approximated by the expression $\sigma = \text{const} \cdot E^m$. It turns out that as the energy interval is increased the parameter m decreases. For the interval $0 < E \leq 12$ eV they found m = 1.02. If the size of the interval approaches zero, then $m \to 1.17$, which is close to Wannier's value. For the ionization of argon the threshold dependence was found to differ from Wannier's result by a larger amount: $\sigma \sim E^{1.3}$.

McGowan and Clarke [96], in their investigation of hydrogen ionization, found that for $E \approx 0.4$ eV the behavior of the cross section agrees with Wannier's expression, in the interval 0.4-3.0 eV it is closest to a linear dependence, and near the threshold ($E \approx 0.05$ eV) the nonlinearity is stronger than Wannier predicts. Their energy resolution was 0.06 eV, so the last result is not completely reliable. However, it should be further noted that in all of these studies [93-96] the exponent is greater than 1.127 in the immediate vicinity of the threshold.

In contrast to the above investigators [93-96], for the ionization of the alkali metals Zapesochnyi and Aleksakhin [97] found that the cross section at threshold depended linearly on energy. This result can be explained by the assumption that the range of nonlinearity for the alkali atoms is very small (~0.1 eV) and, hence, was not observed by Zapesochnyi and Aleksakhin. Indeed, the smaller the binding energy of an electron in an atom, the larger the distance at which the effects of polarization, exchange, and strong coupling manifest themselves. In addition, the higher partial waves play a more important role. From Eq. (24.30), which is based on the similarity principle, it is also seen that as the ionization potential is decreased, the cross section shifts toward lower incident-electron energies, i.e., the interval of nonlinearity is narrowed.

The foregoing investigators [98-97] had studied the threshold behavior of the total ionization cross section. However, of greater interest for comparison with the Wannier theory is the study of the threshold behavior of the differential ionization cross section -- the verification of the existence of a maximum at $\theta = \pi$ and of the uniformity of the distribution over the energies of the outgoing electrons. In recent years coincidence techniques have been developed that allow the simultaneous detection of outgoing electrons. Ehrhardt *et al.* [15,16] studied the distribution of electrons over θ for the ionization of helium, principally at large energies. The lowest value of E was 6 eV. For large E the distribution over θ has two maxima. The so-called binary maximum is situated at an angle θ near 90° and can be interpreted as the result of collisions between free electrons (in the laboratory system particles with identical masses are scattered at an angle of 90°). The second maximum is located at larger θ and corresponds to processes with greater momentum transfer to the nucleus. As E decreases, the role of the second maximum is enhanced. However, for E = 6 eV a well-defined binary maximum is still found, but at $\theta > 90°$. The second maximum at E = 6 eV is probably considerably greater than the binary maximum, but it is in an experimentally inaccessible region (it was not possible to detect electrons departing with an angle >125° from the direction of the incident electron).

The energy and angular dependence of the differential ionization cross section for helium at extremely small energies was investigated by Cvejanović and Read [14]. By means of coincidence techniques they measured the distribution over T -- the difference in the arrival times at the detector of the electrons outgoing from a single ionization event. The basic idea of the experiment was discussed in §3. The measurements were made for electrons departing in opposite directions ($\theta \approx 180°$). The time difference T was determined for $r_1 = r_2$. The distributions of ionization events with respect to T and with respect to the energy of one of the electrons are related by

$$P_T(T) = P_\varepsilon(\varepsilon)/(dT/d\varepsilon) \quad . \tag{28.1}$$

Differentiating (3.28) for $r_1 = r_2 = \text{const}$, we find

$$\frac{dT}{d\varepsilon} = \text{const}\left[\frac{1}{(E-\varepsilon)^{3/2}} + \frac{1}{\varepsilon^{3/2}}\right] \quad . \tag{28.2}$$

Measurements were made for $0.2 \le E \le 0.8$ eV. It turns out that with an accuracy of 15% the measured P_T agrees with Eqs. (28.1) and (28.2), if we set $P_\varepsilon = \text{const}$. Thus, the results corroborate the uniform distribution with respect to the energy of the outgoing electrons.

Cvejanović and Read [14] also investigated the E-dependence of the distribution over θ. For this purpose, at energies in the range $0.2 \le E \le 3.0$ eV, they measured the ratio of the number of ionization events when $\theta = 180° \pm 10°$ to the number of events when $\theta = 150° \pm 10°$. In these measurements there is a larger uncertainty in the experimental results, but they are not in contradiction with the theoretical curve that was obtained under the assumption that the distribution over θ is a Gaussian with width diminishing as $E^{1/4}$.

In addition, they examined the threshold behavior of part of the total ionization cross section; specifically they examined the E-dependence of the number of electrons leaving with very small energies (<0.03 eV). If $d\sigma/d\varepsilon$ is independent of ε, then the dependence of the partial cross section on E is the same as the dependence of the differential cross section. It was found that in the range $0.2 \leq E \leq 1.7$ eV the partial cross section varies as E^m, where $m = 0.131 \pm 0.019$.

The results of [14] also show the existence of a minimum in the differential cross section $d\sigma/d\varepsilon$ at $|E| \approx 0$. A similar result was obtained in [180].

CHAPTER VIII

APPROXIMATE METHODS OF CALCULATING THE EFFECTIVE IONIZATION CROSS SECTION

§29. *The Classical Binary-Encounter Approximation*

Although the classical binary-encounter approximation is the simplest method of calculating effective cross sections, it gives good results in a number of cases. In this approximation the incident electron is considered to interact with only one of the atomic electrons; i.e., it reduces to the scattering of two free electrons, which is described by the Rutherford formula.

This approximation was first considered by Thomson [98], Williams [99], and Thomas [100], but the binary-encounter approximation in the theory of electron-atom collisions was widely applied only after the work of Gryzinski [101].

The binary-encounter approximation takes on a very simple form if we assume that the velocity of the atomic electron is zero before the collision. Then the Rutherford formula can be applied in the laboratory frame, where it has the form [75]

$$\frac{d\sigma}{d\theta_1} = 2\pi \left(\frac{e^2}{E_1}\right)^2 \frac{\cos\theta_1}{\sin^3\theta_1} \quad ; \qquad (29.1)$$

here e is the electron charge, E_1 is the initial kinetic energy, and θ_1 is the scattering angle of the incident electron.

The energy, in the laboratory frame, imparted by the incident electron to the atomic electron is

$$\Delta E = E_1 \sin^2 \theta_1 \quad . \tag{29.2}$$

The Rutherford formula can be rewritten in a form expressing the differential cross section with respect to the energy transfer:

$$\frac{d\sigma}{d(\Delta E)} = \frac{\pi e^4}{E_1 (\Delta E)^2} \quad . \tag{29.3}$$

Ionization corresponds to those scattering events for which $\Delta I \geq I$, where I is the ionization potential. Hence,

$$\sigma = N \int_I^{E_1} \frac{d\sigma}{d(\Delta E)} \, d(\Delta E) \quad , \tag{29.4}$$

where N is the number of electrons in the shell from which the ionization takes place.

Substituting (29.3) into (29.4), we obtain the Thomson equation

$$\sigma = N \frac{\pi e^4}{E_1} \left(\frac{1}{I} - \frac{1}{E_1} \right) \quad . \tag{29.5}$$

This result can be written in the form

$$\sigma = N \pi a_0^2 \left(\frac{Ry}{I} \right)^2 \bar{\sigma}(x) \quad , \tag{29.6}$$

where a_0 is the atomic unit of length (radius of the first Bohr orbit), Ry is the ionization potential of the hydrogen atom, x is the incident-electron energy in threshold units,

$$x = \frac{E_1}{I} \quad , \tag{29.7}$$

and $\bar{\sigma}$ is the reduced ionization cross section,

$$\bar{\sigma}(x) = \frac{4(x-1)}{x^2} \quad . \tag{29.8}$$

The ionization cross section in the form (29.6) satisfies the scaling law expressed by (24.30). The reduced cross section $\bar{\sigma}$ is universal in that it is the same for different initial conditions and different atoms.

The expression for the differential cross section becomes more complicated if the initial velocity of the atomic electron is nonzero. The cross section will depend on the way in which the directions of the atomic-electron velocities are distributed. If this distribution is isotropic, then [102,103]

$$\frac{d\sigma}{d(\Delta E)} = \frac{\pi e^4}{E_1(\Delta E)^2}\sqrt{\frac{E_{min}}{E_2}}\left(1 + \frac{4}{3}\cdot\frac{E_{min}}{|\Delta E|}\right) \quad , \tag{29.9}$$

where E_2 is the initial kinetic energy of the atomic electron and E_{min} denotes the least of the initial and final kinetic energies of the incident and atomic electrons:

$$E_{min} = \min(E_1, E_2, E_1', E_2') \quad . \tag{29.10}$$

If $E_2 = 0$, then (29.9) agrees with (29.3). Equation (29.3) is valid if $E_2 = E_{min}$ and $\Delta E >> E_2$.

In calculations one frequently sets $E_2 = I$, but in the general case the distribution of the initial kinetic energies of the atomic electron can differ from a δ-function distribution. The ionization cross section should then be averaged over E_2.

Setting $E_2 = I$ and substituting (29.9) into (29.4), we can write the result in the form (29.6), where (29.8) is replaced by

$$\bar{\sigma}(x) = \frac{8(x-1)^{3/2}}{3x} \quad , \quad 1 \le x \le 2 \quad , \tag{29.11}$$

$$\bar{\sigma}(x) = \frac{4}{x}\left(\frac{5}{3} - \frac{1}{x-1}\right) \quad , \quad x \ge 2 \quad . \tag{29.12}$$

Gryzinski [101] has introduced an additional simplification. In the derivation of the differential cross section the relative velocity of the electrons $\underset{\sim}{v}_1 - \underset{\sim}{v}_2$ was replaced by its mean value $(v_1^2 + v_2^2)^{1/2}$. Instead of (29.11) and (29.12) the following expressions were obtained:

$$\bar{\sigma}(x) = \frac{16\sqrt{2}}{3x}\left(\frac{x-1}{x+1}\right)^{3/2}, \quad 1 \le x \le 2 , \tag{29.13}$$

$$\bar{\sigma}(x) = \frac{4\sqrt{x}}{(x+1)^{3/2}}\left(\frac{5}{3} - \frac{2}{x}\right), \quad x \ge 2 . \tag{29.14}$$

Here the value of $\bar{\sigma}$ is smaller than in (29.11) and (29.12) and, therefore, agrees better with the experimental data. Equations (29.11) and (29.12) lead to appreciable overestimates of the ionization cross section (see Fig. 10).

The binary-encounter approximation is justified if the effective interaction occurs when the electrons are close together. However, the velocity of the incident electron is greater near the atom than far from it. Because the potential energy of the electron in the atom is approximately equal to 2I, it is reasonable to replace E_1 in Eq. (29.9) by $E_1 + 2I$. We then find with $E_2 = I$ [104]

$$\bar{\sigma}(x) = \frac{4}{3} \cdot \frac{x-1}{x(x+2)}\left(5 + \frac{2}{x}\right), \quad x \ge 1 . \tag{29.15}$$

The ionization cross sections for Mg, Ca, Sr, and Ba [104] calculated from (29.15) turned out to be in good agreement with the Born approximation and with the experimental data. Similar results are found for ionization from the s and p shells. In the ionization of a d electron the binary-encounter cross section is considerably greater than the Born cross section (by a factor as large as 10) [105].

We note that Eqs. (29.12), (29.14), and (29.15) coincide for $x \to \infty$ but differ from (29.8) by a numerical factor.

As $x \to \infty$, the classical binary-encounter cross section falls off as $1/x$, while the quantum cross section falls off as $\ln x/x$. This difference is not due to the use of classical mechanics but rather to the binary-encounter nature of the approximation, as shown by Vriens [106]. In a binary collision both the total energy and the total momentum of the colliding particles are conserved. Therefore, the lower limit of the momentum transfer is larger than when the interaction with the nucleus is taken into account. On the other hand, when the energy of the incident electron is large the region of small momentum transfers is the most important. It should be mentioned in addition that the Rutherford formula, on which the binary-encounter approximation is based, is valid in classical as well as quantum mechanics. Ionization cross sections falling off as $\ln x/x$ can be obtained in the binary-encounter approximation by assuming a particular form for the distribution of initial velocities of the atomic electron, but the distribution function used by Gryzinski [107] in this connection has no physical justification.

A number of authors have investigated the behavior of the total [108-110] and differential [110-113] ionization cross sections for the hydrogen atom in the binary-encounter approximation using the quantum distribution of the atomic-electron initial velocities. The results were compared with the cross sections in the Born approximation. If for a given principal quantum number n the atomic-electron velocity distribution is averaged over the orbital and magnetic quantum numbers ℓ and m, then the quantum distribution for the hydrogen atom is the same as the classical microcanonical distribution. The binary-encounter ionization cross section then satisfies the scaling law that can be written in the form (29.6), where $\bar{\sigma}$ is independent of n.

In Fig. 10 $\bar{\sigma}(x)$ is given for various versions of the binary-encounter approximation. Values of σ corresponding to the experimental data on hydrogen-atom ionization from the 1s level and the cross sections in the Born approximation are also shown.

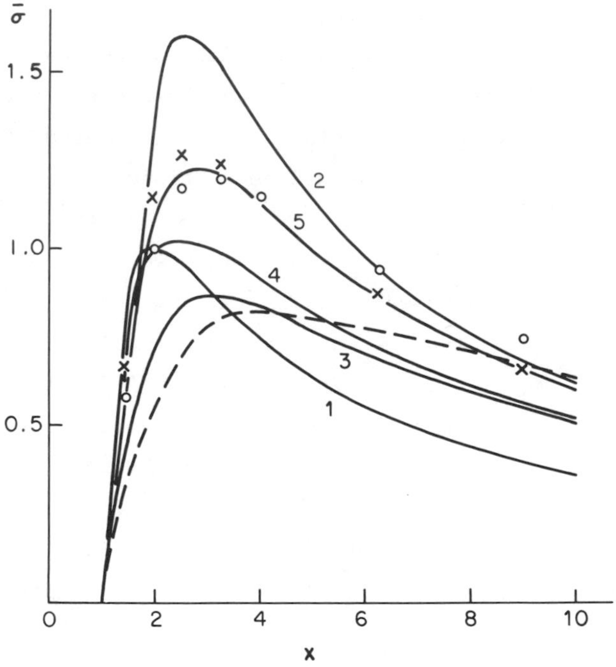

Figure 10. Reduced ionization cross section: 1) from Eq. (29.8); 2) from (29.11) and (29.12); 3) from (29.13) and (29.14); 4) from (29.15); 5) binary-encounter approximation using the quantum distribution of H-atom velocities; ---- is the experimental cross section for H ionization from the 1s level [141]; o is the Born approximation for H ionization from the 1s level; × is the reduced Born cross section for H ionization from the n = 5 level.

The cross section $\bar{\sigma}$ which is obtained when the Born ionization cross section is represented in the form (29.6) depends on n, but not strongly (see Fig. 10). As n increases, the Born cross section $\bar{\sigma}$ approaches the binary-encounter cross section [108-110]. The greatest relative difference between the binary-encounter and Born ionization cross sections occurs near the ionization threshold and for large incident-electron energies. The differential cross section in the binary-encounter approximation agrees well with the Born cross section when the energy transfer ΔE or momentum transfer q is large [110-113]. Bell *et al.* [113] have obtained similar results for helium ionization.

Kingston [109] and Vriens and Bonsen [111] have examined the binary-encounter cross sections for ionization from states with definite quantum numbers n and ℓ (i.e., averaged only over m) using the appropriate quantum distributions of the atomic-electron velocities. In this case, too, the binary-encounter cross section approaches the Born cross section as n increases, and the differential cross sections coincide for ΔE or q sufficiently large.

The quantum velocity distribution, averaged over the quantum number m, is spherically symmetrical. Banks *et al.* [112] have calculated the binary-encounter cross section for ionization of the 2p0 state, for which the velocity distribution is anisotropic. In the binary-encounter differential cross section they found the minimum characteristic of the Born differential cross section for ionization of the 2p0 state.

A number of studies [108-113] indicate that if the quantum distribution of the atomic-electron initial velocities is used, then the binary-encounter approximation can be considered as a simplified variant of the Born approximation.

It has been shown [114,115] that if the Coulomb wave function describing the electron ejected from the atom is replaced by a plane wave in the Born matrix element, the resulting differential cross section (with respect to the energy and momentum transfer) is the same as that in the binary-encounter approximation with the quantum velocity distribution.

The binary-encounter differential cross section has the extremely simple form [100,106,115]

$$\frac{d^2\sigma}{d(\Delta E)dq} = \frac{4\pi e^4}{v_{01}^2 v_{02} q^4} , \tag{29.16}$$

where $\underset{\sim}{v}_{01}$ and $\underset{\sim}{v}_{02}$ are the initial velocities of the incident and atomic electrons. Equation (29.16) is based on an isotropic atomic-electron velocity distribution. In a binary collision the total momentum and energy of the electrons is conserved; hence,

$$\underset{\sim}{v}_{01} + \underset{\sim}{v}_{02} = \underset{\sim}{v}'_1 + \underset{\sim}{v}'_2 , \tag{29.17}$$

$$v_{01}^2 + v_{02}^2 = v_1'^2 + v_2'^2 , \tag{29.18}$$

where $\underset{\sim}{v}'_1$ and $\underset{\sim}{v}'_2$ are the final velocities of the incident and atomic electrons. The momentum and energy transfer are

$$\underset{\sim}{q} = m(\underset{\sim}{v}_{01} - \underset{\sim}{v}'_1) = m(\underset{\sim}{v}'_2 - \underset{\sim}{v}_{02}) , \tag{29.19}$$

$$\Delta E = \frac{1}{2} m(v_{01}^2 - v_1'^2) = \frac{1}{2} m(v_2'^2 - v_{02}^2) , \tag{29.20}$$

where m is the electron mass. Equation (29.19) determines the following range for possible values of q:

$$v_{01} - v'_1 \le q/m \le v_{01} + v'_1 , \quad v_{01} \le v'_2 , \tag{29.21}$$

$$v'_2 - v_{02} \le q/m \le v'_2 + v_{02} , \quad v_{01} \ge v'_2 . \tag{29.22}$$

By integrating (29.16) over q within the limits (29.21) and (29.22), we obtain (29.9).

The equations for the differential cross section in the Born approximation are given in §30. If in the matrix element (30.6) we replace the Coulomb wave function by the normalized plane wave

$$\phi^*(\underset{\sim}{k}_2,\underset{\sim}{r}_2) \to (2\pi)^{-3/2} \exp(-i\underset{\sim}{k}_2 \cdot \underset{\sim}{r}_2) \quad , \tag{29.23}$$

and in Eq. (30.10) transform to an integration over v_{02} by means of the expressions

$$v_{02} = |\underset{\sim}{k}_2 - \underset{\sim}{q}| \ , \qquad d\hat{\Omega}_2 = \frac{2\pi v_{02}}{k_2 q}\, dv_{02} \quad , \tag{29.24}$$

Eq. (30.12) will take the form

$$\frac{d\sigma}{d\varepsilon} = \frac{4\pi}{k_0^2} \iint \frac{P(v_{02})}{v_{02}q^4}\, dv_{02}\, dq \quad , \tag{29.25}$$

where P(v) is the quantum distribution of the absolute values of the atomic-electron velocity:

$$P(v) = \frac{4\pi v^2}{2\ell_0+1} \sum_{m_0} \left| \int \phi_{n_0\ell_0 m_0}(\underset{\sim}{r}_2) \frac{\exp(-i\underset{\sim}{v}\cdot\underset{\sim}{r}_2)}{(2\pi)^{3/2}}\, d\underset{\sim}{r}_2 \right|^2 . \tag{29.26}$$

Because we are using here atomic units in which $|e| = m = 1$, Eq. (29.25) agrees in form with (29.16) averaged over v_{02} and integrated over q; however, the limits of integration are different. In the binary-encounter approximation v_{02} varies from zero to ∞, and the range of variation of q is determined by (29.21) and (29.22). In the Born approximation q varies through the range (30.13), which corresponds to the use of (29.21) for all v_{01} with v_{02} assuming values in the range

$$|k_2 - q| \leq v_{02} \leq k_2 + q \ , \quad k_2^{\ 2} = 2\varepsilon = 2(\Delta E - I) \ . \qquad (29.27)$$

The difference in the ranges of integration is apparent in the behavior of the total ionization cross section at large incident-electron energies. In the binary approximation the cross section falls off as

$$\sigma \sim E_1^{-1} \ . \qquad (29.28)$$

In the simplified Born approximation it goes to a finite limit:

$$\sigma \sim \text{const} \ . \qquad (29.29)$$

The correct behavior of the cross section is given by Eq. (30.23). Thus, neither the binary-encounter nor the simplified Born approximation can be used to calculate the total ionization cross section at large incident-electron energies, because for large E_1 the main contribution to the ionization cross section comes from small momentum transfer q. In the binary-encounter approximation small q are not taken into account because the range of integration over q is too restricted. In the simplified Born approximation the error in the matrix element at small q becomes important. The substitution (29.23) destroys the orthogonality of the initial and final states of the atomic electron, so for small q, when exp(iqr) can be replaced by 1 + iqr, the term containing 1 does not vanish.

The binary-encounter and simplified Born ionization cross sections also differ at threshold, where $\sigma \sim E^{3/2}$ and $\sigma \sim E^2$, respectively, are predicted.

§30. *Quantum-Mechanical Approximations*

Born approximation. At present the Born approximation is the principal method of calculating effective ionization cross sections.

We obtain the ionization amplitude in the Born approximation by substituting the expression

$$\Psi \to e^{i\mathbf{k}_0\cdot\mathbf{r}_1}\phi_0(\mathbf{r}_2) \quad , \tag{30.1}$$

which describes the initial state, for the wave function Ψ in the integral expressions (17.8) and (17.10), and using (17.6) as the final-state wave function. When Ψ is replaced by (30.1) the phase β in Eqs. (17.8) and (17.10) becomes indeterminate. We shall assume that $\beta = 0$. Taking (21.4) into account and using the normalized Coulomb wave function (1.7), we find

$$f(\mathbf{k}_1,\mathbf{k}_2) = \frac{1}{2\pi}\int e^{i\mathbf{k}_0\cdot\mathbf{r}_1}\phi_0(\mathbf{r}_2)\left(\frac{1}{r_1} - \frac{1}{r_{12}}\right)e^{-i\mathbf{k}_1\cdot\mathbf{r}_1}\phi^*(\mathbf{k}_2,\mathbf{r}_2)d\mathbf{r}_1 d\mathbf{r}_2 \, , \tag{30.2}$$

where

$$k_1^{\,2} + k_2^{\,2} = 2E \quad . \tag{30.3}$$

Below, we use atomic units.

The term containing $1/r_1$ in the integral (30.2) vanishes owing to the orthogonality of the atomic wave functions. The second term can be integrated with respect to $\mathbf{r}_1$ by means of the equation [8]

$$\int e^{i\mathbf{q}\cdot\mathbf{r}_1} \frac{1}{r_{12}} d\mathbf{r}_1 = \frac{4\pi}{q^2} e^{i\mathbf{q}\cdot\mathbf{r}_2} \quad . \tag{30.4}$$

We then obtain

$$f(\mathbf{k}_1,\mathbf{k}_2) = -\frac{2}{q^2} \langle 0|\mathbf{q}|\mathbf{k}_2\rangle \quad , \tag{30.5}$$

where

$$\langle 0|\mathbf{q}|\mathbf{k}_2\rangle = \int \phi_0(\mathbf{r}_2) e^{i\mathbf{q}\cdot\mathbf{r}_2} \phi^*(\mathbf{k}_2,\mathbf{r}_2) d\mathbf{r}_2 \quad , \tag{30.6}$$

and $\mathbf{q}$ is the momentum transfer:

$$\mathbf{q} = \mathbf{k}_0 - \mathbf{k}_1 \quad . \tag{30.7}$$

According to (3.24) the differential ionization cross section has the form

$$\frac{d\sigma}{d\varepsilon} = \frac{k_1 k_2}{k_0} \int |f(\mathbf{k}_1,\mathbf{k}_2)|^2 d\hat{\Omega}_1 d\hat{\Omega}_2 \quad , \tag{30.8}$$

where

$$\varepsilon = k_2^2/2 \quad . \tag{30.9}$$

The ionization cross section averaged over the orbital-angular-momentum projection m_0 of the initial state is usually of physical interest. Having been averaged over m_0 and integrated over $\hat{\Omega}_2$, the differential cross section is independent of the direction of $\mathbf{q}$. The expression

$$|\langle n_0\ell_0|q|\varepsilon\rangle|^2 = \frac{k_2}{2\ell_0+1} \sum_{m_0} \int |\langle n_0\ell_0 m_0|\mathbf{q}|\mathbf{k}_2\rangle|^2 d\hat{\Omega}_2 \tag{30.10}$$

determines the angular distribution of the scattered electrons. We use the relation, implied by (30.7),

$$\sin\theta_1\, d\theta_1 = \frac{q\,dq}{k_0 k_1} \quad . \tag{30.11}$$

Integration over ϕ_1 reduces to multiplication by 2π. We find as a result

$$\frac{d\sigma}{d\varepsilon} = \frac{8\pi}{k_0^2}\int_{q_{min}}^{q_{max}} q^{-3}|\langle n_0\ell_0|q|\varepsilon\rangle|^2 dq \quad , \tag{30.12}$$

where

$$q_{min} = k_0 - k_1\,, \qquad q_{max} = k_0 + k_1 \quad . \tag{30.13}$$

If we describe the final state by means of a set of atomic wave functions having definite values of total and component angular momenta,

$$\phi_{\varepsilon\ell m}(\mathbf{r}_2) = R_{\varepsilon\ell}(r_2)Y_{\ell m}(\hat{\Omega}_2) \quad , \tag{30.14}$$

where $R_{\varepsilon\ell}$ is normalized to $\delta(\varepsilon - \varepsilon')$, then Eq. (30.10) takes the form

$$|\langle n_0\ell_0|q|\varepsilon\rangle|^2 = \sum_{\ell=0}^{\infty} |\langle n_0\ell_0|q|\varepsilon\ell\rangle|^2 \quad , \tag{30.15}$$

where

$$|\langle n_0\ell_0|q|\varepsilon\ell\rangle|^2 = \frac{1}{2\ell_0+1}\sum_{m_0}\sum_{m}\left|\int \phi_{n_0\ell_0 m_0}(\mathbf{r}_2)e^{i\mathbf{q}\cdot\mathbf{r}_2}\phi^*_{\varepsilon\ell m}(\mathbf{r}_2)d\mathbf{r}_2\right|^2 . \tag{30.16}$$

Expanding the exponential in partial waves, we can write the result in the form

$$|\langle n_0\ell_0|q|\varepsilon\ell\rangle|^2 = \sum_t B_{\ell_0\ell t}\left|\int_0^{\infty} R_{n\ell_0}(r)j_t(qr)R_{\varepsilon\ell}(r)r^2 dr\right|^2 , \tag{30.17}$$

where j_t is a spherical Bessel function:

$$j_t(x) = \sqrt{\frac{\pi}{2x}}\, J_{t+1/2}(x) \quad , \tag{30.18}$$

$$t = |\ell_0-\ell|\,,\ \ |\ell_0-\ell|+2,\ldots,\ell_0+\ell \quad . \tag{30.19}$$

The equations determining $B_{\ell_0\ell t}$ are given in [4].

For many-electron atoms the ionization amplitude is determined by the matrix element

$$f = -\frac{2}{q^2}\int \phi_0 \sum_a e^{i\underset{\sim}{q}\cdot\underset{\sim}{r}_a}\phi_n^* d\tau_a \quad , \tag{30.20}$$

where the summation and integration are performed over the coordinates of the atomic electrons. If the atomic wave functions are constructed from single-electron wave functions, the expression for the differential cross section in the case of a many-electron atom can also be reduced to expressions of the form (30.12), (30.15), and (30.17). The calculation of the ionization cross section then amounts to the evaluation of integrals containing radial wave functions. For hydrogen it is more convenient to calculate the integral (30.6), which can be evaluated in analytic form. This is true also for many-electron atoms, if comparatively simple analytic functions are used. If the radial functions are given in numerical form, then it is necessary to use (30.15) and (30.17).

Note that the angular distribution of the scattered electrons is often described in terms of the so-called generalized oscillator strength [116],

$$F_{n\ell}(q) = \frac{2(\varepsilon_n-\varepsilon_0)}{q^2}\,|\langle n_0\ell_0|q|n\ell\rangle|^2 \quad . \tag{30.21}$$

The extension of this expression to the continuous spectrum is usually denoted by $dF_{\varepsilon\ell}/d\varepsilon$.

In recent years ionization cross sections in the Born approximation have been calculated for a number of complex atoms, which have been represented by a variety of approximate atomic wave functions.

Ionization cross sections, calculated with the semiempirical wave functions of Vainshtein [117], are given in the monograph of Vainshtein *et al.* [4]. The results are approximated by the analytical expression

$$\sigma = \pi a_0^2 \frac{m}{2\ell_0+1} \left(\frac{Ry}{I}\right)^2 \left(\frac{x-1}{x}\right)^{3/2} \frac{C}{x-1+\phi} , \qquad (30.22)$$

where m is the number of electrons equivalent to the optical electron in the initial state, I is the ionization potential, $x = E_1/I$, and E_1 is the energy of the incident electron. The parameters C and ϕ are tabulated in the monograph [4] for the atoms H-Mg (i.e., for an interval of the periodic table in order of increasing atomic number), as well as for Ar, K, Rb, and Cs.

Vainshtein *et al.* [104] obtain the Born cross sections for the alkaline earth atoms Mg, Ca, Sr, and Ba, which are in good agreement with the experimental data.

Peach has calculated the ionization cross sections for He, Li, Be, Na, and Mg atoms [118] and for atoms with an outer 2p electron (B-Ne) and with an outer 3p electron (Aℓ-Ar) [119]. She has also considered ionization from inner 2s, 2p, and 3s shells [120]. However, owing to an error in the computer program her results for ionization from the 2p and 3p shells [119,120] were too large by a factor of about two. When this error was corrected [121], the results were close to the experimental findings. An appreciable role is played by ionization from the inner shells. Peach [118-121] described the initial states of the atoms by many-parameter analytic Hartree-Fock wave functions, and the final state of the ejected electron by a Coulomb wave function for charge $Z = 1$. The wave functions of the initial and final states were orthogonalized. It should be mentioned that in [118-121] Peach calculated a modified Born approximation, which differed from the usual Born approximation in that the integration over the energy of the atomic electron was restricted to the interval $0 \le \varepsilon \le E/2$. This is correct only when exchange effects are taken into account (§22). In the modified Born approximation the case in which the energy of the ejected electron is greater than the energy of the scattered electron is not taken into account. If such processes result from electron exchange, then the modified approximation is analogous to the Born approximation for the excitation of discrete levels. In the range $E/2 < \varepsilon \le E$ the Born approximation is unsatisfactory, because the faster of the outgoing electrons is described by a Coulomb function and the slower by a plane wave. As the Born cross section is usually greater than the experimental cross section, diminishing the range of integration improves the agreement with experiment. The two versions of the Born approximation are similar for large incident-electron energies, for which the interval $E/2 < \varepsilon \le E$ makes a small contribution to the ionization cross section.

The ionization of the hydrogen atom from the 2s and 2p states was investigated by Kyle and Omidvar [122].

The ionization cross sections for He, Li, C, N, O, Ne, Na, Mg, Ar, K, and Zn were calculated by Omidvar *et al.* [123]. For the initial state they used hydrogen-like wave functions, the effective charges of which were chosen from the results of the Hartree-Fock method, and for the final state they used the Coulomb functions corresponding to $Z = 1$. The calculated cross sections turn out to be similar to those of Peach [121], but the calculation is less complicated. Omidvar *et al.* [123] take into account the ionization of s, p, and d electrons of the atom. The total ionization cross sections are compared with experiment; for He and N the differential cross sections are compared as well. For atoms with small atomic numbers the results are close to the experimental ones, but as the atomic number increases the theoretical cross sections exceed the experimental ones.

McGuire [124] has obtained the ionization cross sections for the sequence of atoms He-Na, allowing for all occupied shells, and the cross section for the ionization of Ar from the 3s and 3p shells. The single-electron wave functions were determined with a simplified form of the Herman-Skillman potential [rV(r) was replaced by line segments]. For $E_1 > 200$ eV McGuire's results differ from the experimental ionization cross sections by less than 20%.

For large incident-electron energies the ionization cross section in the Born approximation can be represented in the form

$$\sigma \sim A\frac{\ln E_1}{E_1} + \frac{B}{E_1} + \frac{C}{E_1^2} + \ldots \quad , \tag{30.23}$$

where A, B, and C are constants. The logarithmic term appears because of the q^{-3} factor, which causes the main contribution to the integral (30.12) to come from the region of small q. This is especially noticeable as $k_0 \to \infty$, in which case

$$q_{min} = k_0 - \sqrt{k_0^2 - 2\varepsilon - 2I} \to 0 \quad . \tag{30.24}$$

For small q the exponential exp(iqr) can be replaced by 1 + iqr. The first term in the resulting expression vanishes owing to the orthogonality of the atomic wave functions, and the second term leads to the logarithmic term in (30.23).

An expansion of the form (30.23) holds also for the differential cross section $d\sigma/d\varepsilon$ (with the coefficients depending on ε) as well as for the total (elastic + inelastic) scattering cross section. The expansion coefficients for the total cross section can be found by means of the sum rules for the generalized oscillator strengths [8,116,125]. Moreover, to determine A and C, it suffices to know the atomic wave function of the initial state, and to find B it is necessary to know the distribution of oscillator strengths [126,127]. By extracting from the total cross section the part pertaining to the discrete levels, Inokuti and Kim [126,127] obtained the coefficients A, B, and C for the ionization cross sections of H, He^+, H^-, He, and Li^+. Omidvar [128] has obtained extremely accurate values of A and B for the ionization of the hydrogen atom by direct series expansion of the analytic expression for the differential ionization cross section.

Bell and Kingston [129] have obtained the coefficients in the expansion (30.23) for the ionization cross section of helium, including terms containing E_1^{-3} and E_1^{-4}, by fitting to the numerical values of an ionization cross section that was calculated with comparatively precise atomic wave functions. For the initial state they used a six-parameter function taking into account electron correlations, and they determined the final-state wave function by the method of polarized orbitals. In addition to the matrix element in the form (30.20), Bell and Kingston use another form to which (30.20) may be reduced if ϕ_0 and ϕ_n are exact solutions of the Schrödinger equation. The different forms of the matrix element are analogous to the corresponding forms used in the calculation of oscillator strengths [130]. The use of their alternative form increases the ionization cross section in the vicinity of the maximum by 10%.

Balashov *et al.* [131] have investigated the angular and energy distributions of the electrons ejected from the atom in the ionization of helium. For the initial state they used the Hartree-Fock wave function and for the final state, the Coulomb wave function for Z = 1. Comparison with the experimental data [15,16,132] for the shape of the spectrum of ejected electrons shows that the Born approximation is more satisfactory than the binary-encounter approximation.

Stingl [133] has studied the effect of the superposition of configurations on the angular distribution of the scattered and ejected electrons in the ionization of helium. It turns out that allowing for the superposition of configurations in the initial-state wave function of the atom leads to small changes (4-8%) in the differential cross section.

Stingl [134] also calculates the ionization cross sections for the boron atom and for a number of isoelectronic ions.

Tweed [135] has considered double ionization of helium by electron impact.

A general formulation in the Born approximation for the differential cross section, in L-S coupling, for complex atoms is given by Robb *et al.* [181].

Exchange effects. The generalization of the Born approximation to account for electron exchange is known as the Born-Oppenheimer approximation. We obtain the ionization amplitude in this approximation if we antisymmetrize the initial or final-state wave function in the integral (30.2). The amplitude f is then replaced by $f \pm g$, where

$$g(\mathbf{k}_1,\mathbf{k}_2) = \frac{1}{2\pi}\int e^{i\mathbf{k}_0\cdot\mathbf{r}_1}\phi_0(\mathbf{r}_2)\left(\frac{1}{r_1} - \frac{1}{r_{12}}\right)e^{-i\mathbf{k}_1\cdot\mathbf{r}_2}\phi^*(\mathbf{k}_2,\mathbf{r}_1)d\mathbf{r}_1 d\mathbf{r}_2 \;. \tag{30.25}$$

The ionization cross section is found from Eq. (22.8). The Born–Oppenheimer approximation usually leads to greatly overestimated cross sections because of errors in expression (30.25) for the amplitude. An obvious deficiency of Eq. (30.25) is that the initial and final states are nonorthogonal. Even in the absence of any interaction between the electrons the amplitude g nevertheless turns out to be nonzero. Thus, Eq. (30.25) for g is unsatisfactory.

However, because of Eqs. (2.3) and (21.8), in the case of ionization there is no need to determine the amplitude g independently of f. Exchange effects can be taken into account in the framework of the Born approximation by using (22.9) and (22.10). In this approach, however, the phase τ [or the phase Δ, in terms of which τ is expressed by Eq. (21.9)] remains unknown. The following expression for Δ has been suggested by Peterkop [136]:

$$\Delta(\underset{\sim}{k}_1,\underset{\sim}{k}_2) = \arg \Gamma(1 - i/k_2) \quad . \tag{30.26}$$

The form (30.26) is convenient in that it compensates for the phase factor in the normalizing factor of the Coulomb function $\phi^*(\underset{\sim}{k}_2,\underset{\sim}{r}_2)$, which enters into (30.2). The use of (30.26) is equivalent to the following procedure. We assume $\Delta = 0$, but subtract $\arg \Gamma(1 - i/k_2)$ from the phase of the amplitude f, or add to it $\arg \Gamma(1 - i/k_1)$. In both cases the amplitude becomes more nearly symmetric with respect to the electrons. In the form (30.2) the amplitude contains a rapidly oscillating factor for small k_2 but not for small k_1.

In order to establish the maximum possible magnitude of the interference effect, we have also calculated [136] the integral (22.10) of the product of the moduli of the amplitudes. Formally, this variant corresponds to

$$\Delta(\underset{\sim}{k}_1, \underset{\sim}{k}_2) = \arg f(\underset{\sim}{k}_1, \underset{\sim}{k}_2) \quad . \tag{30.27}$$

The results, shown in Fig. 11, indicate that allowing for interference does improve considerably the agreement of the Born approximation with experiment. Geltman *et al.* [137] have calculated a version with $\Delta = 0$. Such a choice for Δ diminishes the interference somewhat and correspondingly enhances the ionization cross section at small energies.

The Born approximation with allowance for exchange has been used in calculations of the ionization cross sections of He by Peach [118] and Sloan [138], of Li, Be, Na and Mg by Peach [118], and in calculations of the ionization of the

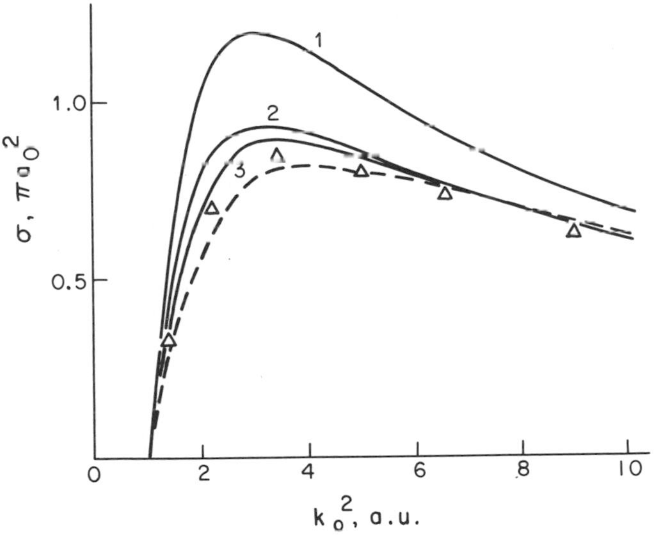

Figure 11. Ionization cross sections for hydrogen: 1) Born approximation; 2) Born approximation allowing for exchange with the choice of phase in the form (30.26); 3) Born-Ochkur approximation using (30.31); — — — results of experiment [141]; Δ - the Born approximation allowing for maximal interference.

hydrogen atom from the 2s and 2p states by Kyle and Omidvar [122], Rudge and Schwartz [139], and Prasad [140]. In most of this work the phase Δ was assumed to have the form (30.26). For helium, allowance for interference reduced the difference between the Born approximation and experiment by at least a factor of two. For heavier atoms and also for ionization from excited states the role of interference diminishes. In the majority of cases allowance for interference will decrease the ionization cross section. An exception to this is the ionization of the hydrogen atom from the 2pm state ($m = \pm 1$), where the cross section is increased by taking interference into account [122].

The Born approximation is unsatisfactory in the energy range $E/2 < \varepsilon \leq E$ ($\varepsilon \equiv \varepsilon_2$), where because $k_2 > k_1$ the faster electron is described in the integral expression (30.2) by the Coulomb wave function and the slower one by the plane wave. But the amplitude g for $\varepsilon < E/2$ is determined by the amplitude f for $\varepsilon > E/2$ by means of Eq. (21.8). In addition, in calculating the interference integral (22.10) with the Born amplitude f, it is necessary to integrate numerically over three of the four angles determining the directions of the outgoing electrons. In the calculation of the Born cross section ignoring exchange effects it is necessary to integrate numerically only over the angle θ_1, which reduces to integration over q in (30.12). The integrals over the remaining angles are evaluated analytically.

A successful improvement in the Born-Oppenheimer method was proposed by Ochkur [142,143]. Because the Born-Oppenheimer approximation is well founded only for large incident-electron energies, Ochkur replaced the amplitude g by the leading term of its expansion in powers of $1/k_0$.

Let us consider the integral over $\mathbf{r}_1$ in (30.25). If k_0 and k_1 are large, then the integrand oscillates rapidly. The main contribution to the integral comes from the vicinity of the singular points $r_1 = 0$ and $\mathbf{r}_1 = \mathbf{r}_2$. We represent the Coulomb function $\phi^*(\mathbf{k}_2,\mathbf{r}_1)$ in the form (1.7), remove the hypergeometric function at the singular points from under the integral sign, and then use Eq. (30.4). The term thus produced at the point $r_1 = 0$ on further integration over $\mathbf{r}_2$ turns out to be of higher order in the small quantity $1/k_0$. Therefore we have, as a result [143],

$$g(\mathbf{k}_1,\mathbf{k}_2) = -\frac{2}{|\mathbf{k}_0-\mathbf{k}_2|^2}\langle 0|\mathbf{q}|\mathbf{k}_2\rangle \quad . \tag{30.28}$$

This expression differs from the amplitude f only by a simple factor. Ochkur [143] suggests further that (30.28) be replaced by the similar, but considerably simpler, expression

$$g(\mathbf{k}_1,\mathbf{k}_2) = -\frac{2}{k_0^2 - k_2^2}\langle 0|\mathbf{q}|\mathbf{k}_2\rangle \quad . \tag{30.29}$$

For the differential ionization cross section averaged over spin we then obtain instead of (30.12)

$$\frac{d\sigma}{d\varepsilon} = \frac{8\pi}{k_0^2}\int_{q_{min}}^{q_{max}} q^{-3} X\,|\langle 0|q|\varepsilon\rangle|^2 dq \quad , \tag{30.30}$$

where

$$X = 1 - \frac{q^2}{k_0^2 - k_2^2} + \left(\frac{q^2}{k_0^2 - k_2^2}\right)^2 \quad . \tag{30.31}$$

The ionization cross section for hydrogen, calculated by Ochkur [143] from (30.30) and (30.31) is close to the experimental cross section (see Fig. 11). Moreover the calculation is even less laborious than in the Born approximation, because the integration over ε is restricted to the interval $0 \leq \varepsilon \leq E/2$.

The Born-Ochkur method has been used by numerous authors [118-122,140] to calculate the ionization cross sections of atoms from H to Ar. Instead of (30.31) they used a variation analogous to that suggested by Ochkur [142] for excitation of discrete levels, in which

$$X = 1 - \frac{q^2}{k_0^2} + \frac{q^4}{k_0^4} \quad . \tag{30.32}$$

Taking the exchange effects into account by the Ochkur method noticeably reduces the Born cross section, which thus improves the agreement with experiment. For the ionization of hydrogen from the 2s and 2p excited states the Ochkur method gives results similar to the calculations in which exchange, allowing for maximum interference, is included in the Born approximation [122]. We expect this to hold also in the ionization of the valence electrons of complex atoms.

Systematically retaining only the leading terms, we may replace Eq. (30.32) by Ochkur's expression [144]

$$X = 1 - \frac{q^2}{k_0^2} \quad . \tag{30.33}$$

In so doing we should set $q_{max} = k_0$. For the ionization of the hydrogen atom the use of (30.33) leads to good agreement with experiment. In this version of the calculation the exchange effects definitely decrease the ionization cross section because $X < 1$; in addition, allowance for exchange reduces the range of integration over ε.

It was shown by Rudge [7] that Eq. (30.28) follows from the variational principle with a special choice of the trial function.

Method of unseparated variables. In the above methods the variables are separated in the sense that the wave function is represented as a product of single-electron wave functions. Vainshtein *et al.* [145] and Presnyakov [146,147] have developed a method in which the wave function in the integral expression for the ionization amplitude takes into account the repulsion between the incident and optical electrons.

For the ionization amplitude they use the expression

$$f = -\frac{1}{2\pi}\int G(\mathbf{r}_1,\mathbf{r}_2)\phi_0(\mathbf{r}_2)\frac{1}{r_{12}}e^{-i\mathbf{k}_1\cdot\mathbf{r}_1}\phi^*(\mathbf{k}_2,\mathbf{r}_2)d\mathbf{r}_1 d\mathbf{r}_2 \quad , \tag{30.34}$$

which follows from (30.2) if the incident plane wave is replaced by G and the term containing $1/r_1$ is omitted. When exchange is taken into account, the initial or final state is appropriately symmetrized. The function G is replaced by a product of Coulomb functions that represents the relative motion of the electrons with velocity $\mathbf{k}_0$ and the motion of the center of mass of the electrons with velocity $\mathbf{k}_0/2$. The effective charge determining the Coulomb motion is found by considering the asymptotic behavior of G and turns out to depend on k_0:

$$\zeta = k_0\nu \quad , \quad \nu = \frac{1}{k_0 + \sqrt{2I}} \quad , \tag{30.35}$$

where I is the ionization potential.

In calculating the integral (30.34) Vainshtein *et al.* [145] introduce several simplifications. They then obtain for the ionization cross section an equation of the form (30.30), where

$$X = \frac{1}{4}X^+ + \frac{3}{4}X^- \quad , \tag{30.36}$$

$$X^{\pm} = \left[y(\nu,x) \pm \frac{q^2}{k_0^2}\, y\left(\nu\, , \frac{1}{4}\right) \right]^2 \quad , \qquad (30.37)$$

$$y(\nu,x) = \frac{\pi\nu}{\operatorname{sh}\pi\nu}\ {}_2F_1(-i\nu,i\nu,1,x) \quad , \qquad (30.38)$$

$$x = \left[\frac{k_0^{\,2} - k_1^{\,2} + q^2}{k_0^{\,2} - k_1^{\,2} + 3q^2} \right]^2 \quad . \qquad (30.39)$$

Evaluation of Eqs. (30.36)-(30.39) [146] gave good results for the ionization of the hydrogen atom, but for sodium the cross section was severely underestimated. It has been shown [122,148] that the simplifications made in the evaluation of the integral (30.34) have had an appreciable effect on the result. The Eqs. (30.37)-(30.39) are, therefore, to a certain extent semiempirical in nature.

Distortion effects. Equations (30.2) and (30.25) for the ionization amplitude were obtained by substituting the product of a plane wave and an atomic wave function for the exact solution of the Schrödinger equation in the integral expressions (17.8) and (17.10). It is possible to improve the Born and Born-Oppenheimer approximations by using instead of the plane wave $\exp(i\mathbf{k}_0 \cdot \mathbf{r}_1)$ a function that takes into account the distortion of the incident wave by the field of the atom -- in other words, a function which describes the elastic scattering of the electron by the atom. Such a function is constructed by the method of partial waves and can be represented in the form

$$F(\mathbf{k}_0,\mathbf{r}_1) = \frac{1}{k_0}\sqrt{\frac{\pi}{2}} \sum_{\ell=0}^{\infty} i^{\ell}(2\ell+1)e^{i\delta_\ell} R_\ell(r_1) P_\ell(\cos\theta_{\mathbf{k}_0 \cdot \mathbf{r}_1}) \quad , \qquad (30.40)$$

where δ_ℓ is the scattering phase shift, and R_ℓ is the radial wave function of the electron in the field of the atom, which is determined by

$$\left[\frac{d^2}{dr^2} - \frac{\ell(\ell+1)}{r^2} - 2(V_s + V_p + W_\ell) + k_0{}^2\right](rR_\ell) = 0 \quad , \qquad (30.41)$$

$$R_\ell(\infty) \sim \sqrt{\frac{2}{\pi}} \cdot \frac{1}{r} \sin\left(k_0 r - \frac{\ell\pi}{2} + \delta_\ell\right) \quad . \qquad (30.42)$$

Here V_s is the static potential, V_p the polarization potential, and W_ℓ the exchange potential. The latter is an integral operator.

The radial functions are usually determined by numerical integration. Further analytical integration over r_1 is then not possible. Therefore, the remaining functions in the integrals (30.2) and (30.25), which describe the states of the incident and atomic electrons and the interaction potential, should also be expanded in a series of angular functions. As a result the calculation of $d\sigma/d\varepsilon$ reduces to a calculation of the matrix elements of the radial functions and a summation over the angular momenta of the incident and atomic electrons and of the potential.

The role of distortion in the ionization of hydrogen was examined by Veldre and Vinkaln [149,150]. Their results are shown in Fig. 12. The versions with exchange taken into account are represented by the curves 3-5. Curve 3 is calculated taking into account V_s and W_ℓ in (30.41). Curves 4 and 5 are obtained by also accounting for V_p in various forms. It is clear that the polarization potential has little effect. On the other hand, allowing for the static and exchange interactions yields a result considerably more accurate than does the Born-Oppenheimer approximation. In comparison with the Born approximation the difference from experiment is reduced by a factor of

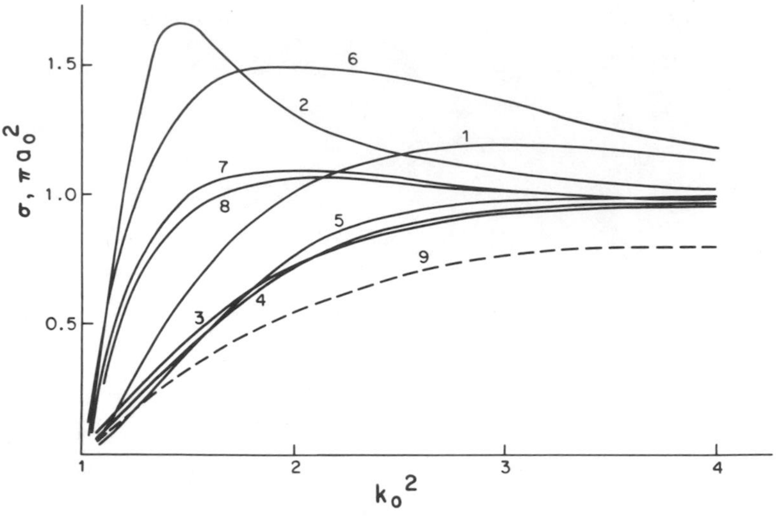

Figure 12. Role of distortion in the ionization of hydrogen: 1) Born approximation; 2) Born-Oppenheimer approximation; 3) Born-Oppenheimer approximation allowing for distortion of the incident wave by the static and exchange potentials; 4,5) the same allowing for the static and for the exchange and polarization potentials; 6) Geltman approximation ignoring exchange; 7) Geltman approximation with exchange; 8) Geltman approximation with exchange and distortion of the incident wave by the static and exchange potentials; 9) experiment [141].

two. The success when distortion is allowed for seems to result from the exchange effects included with the distortion. This conclusion follows from a comparison with the calculations of the excitation of the 2s and 2p levels, in which the distorted-wave method with no exchange [151] gives a considerably overestimated cross section, while inclusion of exchange effects [152] leads to satisfactory results.

Veldre and Vinkaln [149,150] have also investigated the influence of the choice for the final-state wave function. In the case of ionization the distorted-wave method cannot be applied to the final state, because the atomic electron in a continuum state does not generate a static potential. Veldre and Vinkaln considered the approach, applied earlier by Geltman [68], in which both electrons in the final state are described by Coulomb functions. The initial state was not changed. In Fig. 12 the initial states used in the calculations 6, 7, 8 are the same as those of curves 1, 2, 3. With the final state in the Geltman form, exchange plays a large role, and the effect of distortion is insignificant.

Liepin'sh and Rabik [153-155] have investigated the influence of initial-wave distortion and the choice of the atomic-electron wave functions in the ionization of the alkali metal atoms Li, Na, K, and Cs. They calculated the partial-wave cross sections for values of the total orbital angular momentum $L = 0,1,\ldots,6$. For the lithium atom they considered $L = 0,1,\ldots,11$ and also calculated the total Born cross section using hydrogen-like functions [154]. Exchange was ignored in these calculations. For the alkali atoms the role of exchange may be less important because the higher partial waves make a larger contribution. The initial states of the atomic electron were described by Hartree-Fock and semiempirical wave functions. The wave function of the electron ejected from the atom was calculated with the potential V^+ used to calculate the initial-state atomic wave function. The distortion of the incident wave was determined with the static and polarization potentials V_s and V_p of the atom. The motion of the incident electron after ionization was described by a free-particle radial function.

For the lithium atom the static potential V_s has comparatively little influence on the ionization cross section, whereas the polarization potential V_p has a greater effect. For Na, K, and Cs the situation is reversed. The main variation of the cross sections is produced by V_s, and the additional inclusion of V_p is less significant. For all of the alkaline atoms considered the allowance for $V_s + V_p$ reduces the sum of the first seven partial-wave cross sections in the region of the maximum by 20-25%. The results of Liepin'sh [154] indicate that the choice of the wave function of the ejected electron is of substantial importance. Replacing the function calculated with $V^+(r)$ by a hydrogen function decreases the sum of the partial-wave cross sections for $L = 0,1,\ldots,6$ by 20%.

Liepin'sh and Rabik [155] have investigated the cross sections for ionization of 6s and 5p electrons in cesium for $L = 0,1,\ldots,6$. Four variations of the calculation were performed: the incident electron was described by a free-particle radial function or by a distorted wave, and the ejected electron was described by an atomic function, calculated with the static potential of the ion formed, or by a free-particle function. When the ejected electron was described by a free-particle radial function, the cross section for ionization of a 6s electron was enhanced by a factor of three. On the other hand, the greatest cross section for a 5p electron was obtained in the usual Born approach, where the incident electron is described by a free-particle wave function and the ejected electron by an atomic wave function.

In connection with the discussion of the role played by the choice of wave function we should mention the work of Rudge and Schwartz [139], who examined the approach in which the initial state of the incident electron is described by a plane wave and the final state by a Coulomb function satisfying (17.18), where $Z_2 = 1$. In the vicinity of the maximum the cross section is overestimated, which is apparently connected with the nonorthogonality of the incident-electron initial and final states.

Inclusion of exchange in this approach increases the cross section considerably.

The most complete treatment of distortion of the initial-state wave function is given by Burke and Taylor [156]. As the initial-state wave function they used a function describing an e-H collision taking into account exchange and strong 1s-2s-2p coupling. Thus, Burke and Taylor have considered the distortion of the initial state of both the incident and the atomic electron. The final state of the electrons was described in the same way as in the Born-Oppenheimer approximation. Recall that in the integrals (17.8) and (17.10) it is not necessary that the final-state wave function be a solution of the Schrödinger equation. The use of an extremely accurate initial-state wave function did not lead to any essential improvement in the results. The difference between the Born approximation and experiment in the vicinity of the maximum was decreased by 1/3.

The Glauber method. McGuire and coworkers have applied to the ionization of hydrogen the Glauber method [157,158], which is a semiclassical correction to the Born method. The incident wave $\exp(i\mathbf{k}_0 \cdot \mathbf{r}_1)$ in the matrix element (30.2) is replaced by the function

$$\exp\left[i\mathbf{k}_0 \cdot \mathbf{r}_1 - \frac{i}{k_0} \int_{-\infty}^{z_1} \left(\frac{1}{r_{12}} - \frac{1}{r_1}\right) dz_1\right] , \qquad (30.43)$$

which allows for the phase change of a fast incident electron due to its interaction with the atom. The integration over the coordinate z_1 is carried out along a straight line parallel to the z_1 axis. If the z_1 axis is perpendicular to the momentum transfer $\mathbf{q}$, then in the matrix element (30.2) the integral over z_1 can be evaluated analytically. The integral over the remaining variables is expressed in terms of an infinite sum of hypergeometric functions. The differential ionization cross section for small q turns out to be similar to the Born cross section; for large q it is considerably greater than the Born cross section.

A more refined closed-form expression for the ionization amplitude in the Glauber approximation is given by Jain and Srivastava [182].

The pseudostate method. The original application of the so-called pseudostates to the calculation of ionization cross sections was proposed by Gallaher [159]. In the general case, an arbitrary state, which in the close-coupling approximation can be used to represent the actual state of the atom, e.g., to take into account polarization, is called a pseudostate [160].

Gallaher [159] considered the collision of an electron with a hydrogen atom. In expanding the wave function in a series of atomic wave functions he replaced the 2s and 2p states with pseudostates $\overline{2s}$ and $\overline{2p}$ which contain radial functions of the type $f(r)e^{-r}$, where f is a polynomial. The calculation was performed in the close-coupling approximation accounting for exchange and $1s$-$\overline{2s}$-$\overline{2p}$ coupling. Cross sections for the excitation of the $\overline{2s}$ and $\overline{2p}$ pseudostates were obtained. As the pseudostates are not eigenstates of the atom, they contain an admixture of all atomic states (with the same orbital angular momentum), including states of the continuous spectrum. That part of the excitation cross section of the pseudostates which is proportional to the contribution from the continuous spectrum can be related to ionization. The ionization cross section was found by Gallaher [159] by subtracting the part of the cross section corresponding to the discrete spectrum:

$$\sigma_{\text{ion}} = \left(1 - \sum_{n=1}^{\infty} |\langle\overline{2s}|ns\rangle|^2\right)\sigma_{\overline{2s}} + \left(1 - \sum_{n=2}^{\infty} |\langle\overline{2p}|np\rangle|^2\right)\sigma_{\overline{2p}} \ . \tag{30.44}$$

For $k_0^2 < 2$ a.u. the calculated results are close to experimental values [141], but for larger k_0^2 they are appreciably too small. This is to be expected, because the pseudostates for $n > 2$ must also contribute to σ_{ion}. Equation (30.44) is deficient in that it can give a nonzero ionization cross section for incident-electron energies below the ionization threshold.

Ionization through autoionizing states. The process of ionization in complex atoms can proceed, in part, through the excitation of autoionizing states which decay with the ejection of an electron. Interference between direct ionization and ionization through autoionizing states produces resonance phenomena in the differential ionization cross section.

The role of autoionizing states in the ionization of helium has been studied by a number of authors [161-166]. Balashov *et al.* [167,168] have used wave functions for autoionizing states that had been determined previously by the diagonalization method (variational method) with the basis functions

$$\Phi_{n\ell n'\ell'} = \phi_{n\ell}(\underset{\sim}{r}_1)\phi_{n'\ell'}(\underset{\sim}{r}_2) \pm (\underset{\sim}{r}_1 \rightleftarrows \underset{\sim}{r}_2) \ , \quad n,n' \geq 2 \ , \tag{30.45}$$

where $\phi_{n\ell}$ is an eigenfunction of the He^+ ion. Using up to ten basis functions, they obtained the wave functions, positions, and widths of the $^{1,3}S^{(+)}$, $^{1,3}P^{(-)}$, and $^{1,3}D^{(+)}$ autoionizing states (the ± sign in the index indicates the parity).

Lipovetskii *et al.* [161] have examined the sensitivity of the full differential cross sections for the excitation of

$^1P^{(-)}$ autoionizing states of helium to the form of the atomic wave functions. Inclusion of electron correlations in the atomic wave function diminishes the cross section considerably. The total excitation cross sections for autoionizing states are very small ($\sigma < 0.001\ \pi a_0^2$) in comparison with the total ionization cross section for helium ($\sigma \sim 0.6\ \pi a_0^2$), because excitation of autoionizing states is a second-order process (simultaneous excitation of both electrons of the atom). The autoionizing states do not affect the total ionization cross section of helium but, owing to the resonance phenomena, can lead to experimentally observable effects in the spectra of the scattered and ejected electrons. The interference phenomena have been studied by numerous authors [162-166]. The effects associated with the state $(2s2p)^1P$ have been studied in the most detail. Other $^1P^{(-)}$ states have also been investigated, as well as $^1S^{(+)}$ and $^1D^{(+)}$ states. The transition matrix elements have been determined in the first Born approximation. Lipovetskii and Senashenko [164] have estimated the role of the terms corresponding to the second Born approximation, and Pavlichenko and Senashenko [165] have taken account of exchange using the Ochkur method.

The differential scattering cross section for incident electrons can be represented in the form [169]

$$\frac{d^2\sigma}{d\hat{\Omega}_1 dE} = \left(\frac{d^2\sigma}{d\hat{\Omega}_1 dE}\right)_{bg} + \left(\frac{d^2\sigma}{d\hat{\Omega}_1 dE}\right)_{res} \frac{[x+y(q)]^2}{x^2+1} , \qquad (30.46)$$

where

$$x = 2(E - E_{res})/\Gamma , \qquad (30.47)$$

the subscripts "bg" and "res" designate "background" and "resonance," respectively. Here Γ is the width of the autoionization level, and y is the profile parameter, defined by Fano [169]. The angular and energy distributions of the ejected electrons can be conveniently written in the form [163]

$$\frac{d^2\sigma}{d\hat{\Omega}_2 dE} = f(\underset{\sim}{k}_2) + \frac{a(\underset{\sim}{k}_2)x + b(\underset{\sim}{k}_2)}{x^2+1} \quad . \tag{30.48}$$

From the calculations it appears that f, a, and b are complicated (non-monotonic) functions of the angle θ_2. Therefore, because of the interference of the resonance with the direct transition to the continuous spectrum, the shape of the resonance depends strongly on the ejection angle. If (30.48) is brought to the form (30.46), the profile parameter $\tilde{y}(\theta_2)$ is also a non-monotonic function of θ_2. The profile parameters of the scattered and ejected electrons can behave differently. The earlier results [162-165] are in satisfactory agreement with the experimental data on the shape of the resonances.

Balashov *et al.* [166] have considered the possibility of studying autoionizing states by coincidence methods (the detection of the electrons associated with a single act of ionization). Coincidence methods are of interest in connection with the possibility of choosing configurations in which the background of direct transitions is less pronounced or in which the superposition of different autoionizing states is absent.

Liepin'sh and Rabik [170] have studied the autoionizing states of cesium. They have obtained the wave functions and energies for the terms of the $5p^5 6sn\ell$ configurations, where $n\ell$ takes the values 6s, 6p, 5d, and 7s. The calculations were performed by the Hartree method, and the 5p, 6s, and $n\ell$ electrons were treated self-consistently. The wave functions obtained in this way were then used [171] to calculate the excitation cross sections for autoionizing states. We note that in contrast with helium, in cesium the excitation of autoionizing states like $5p^5 6snℓ$ is a one-electron process.

Liepin'sh and Rabik [171], in their main calculations, used both the Born-Oppenheimer approximation and a procedure that allowed for distortion of the incident wave. They took into account nine partial waves, L = 0,1,...,8. The results of the calculations are shown in Fig. 13. The excitation cross section of the $5p^5 6s5d$ state is the largest and is comparable with the direct ionization cross section. Allowance for distortion of the incident wave decreases considerably the excitation cross section. Liepin'sh and Rabik did not consider interference effects.

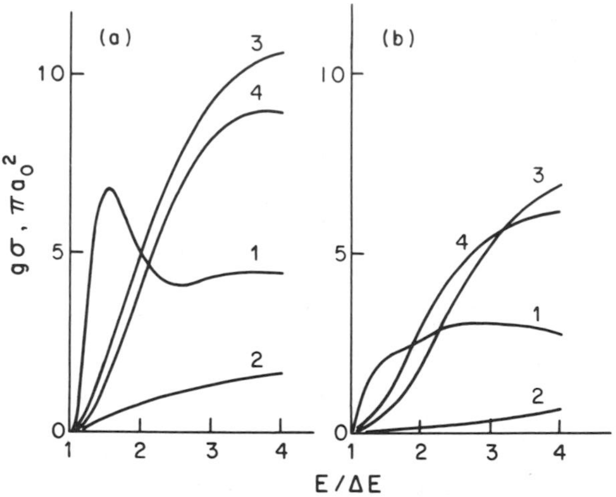

Figure 13. Excitation cross sections of autoionizing states in cesium: a) Born-Oppenheimer approximation; b) allowance for incident-wave distortion; 1) $5p^5 6s^2$; 2) $5p^5 6s6p$; 3) $5p^5 6s5d$; 4) $5p^5 6s7s$. The scale factors are $g_1 = 10$, $g_2 = g_3 = 1$, $g_4 = 100$. The incident-electron energy is expressed in threshold units.

Figure 14 shows the ionization cross sections for cesium taking into account autoionizing states. The ionization cross sections for 6s and 5p electrons were obtained by Liepin'sh and Rabik [155] for L = 0,1,...,6, which is perfectly adequate (at the energies in question) for the ionization of a 5p electron but is insufficient for the ionization of a 6s electron. Only

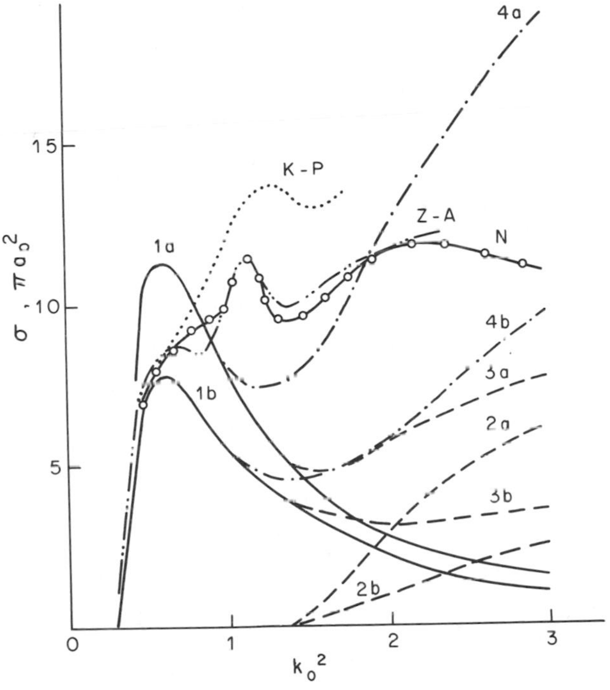

Figure 14. Ionization cross sections for cesium: a) Born approximation (Born-Oppenheimer approximation for autoionizing states); b) allowance for incident-wave distortion; 1) ionization cross section for a 6s electron (with L = 0,1,...,6); 2) ionization cross section for a 5p electron; 3) overall ionization cross section for 6s and 5p electrons; 4) overall ionization cross section for 6s and 5p electrons and excitation of autoionizing states; K-P, Z-A and N refer to the experimental results [172], [97] and [173], respectively.

the shape of the cross section can be compared with experiment. The experimental cross section has several maxima. The theoretical cross section has two maxima, the first of which is due to the ionization of a 6s electron, while the second arises from the ionization of a 5p electron and autoionization. The difference between the curves 4a and 4b in Fig. 14 shows that a large role is played by the distortion of the incident wave.

Autoionizing states of Na, K, and Rb atoms and their excitation cross sections were studied by Rabik [177], who also gives detailed results for Cs.

Experimental ionization cross sections for K, Rb and Cs have also been obtained by Nygaard [183]. These results are close to those in [97]. The effects of autoionization and inner-shell ionization appear to be smaller than predicted by binary-encounter calculations.

Cross sections for the excitation of the lowest autoionizing levels of K, Rb and Cs in the Born and Vainshtein approximations have been calculated by Tiwary and Rai [184] using a variety of atomic wave functions. Their results are significantly larger than those obtained in [171] and [177].

REFERENCES

1. G. F. Drukarev, *Theory of Electron-Atom Collisions* [in Russian], Fizmatgiz, Moscow (1963); trans: S. Chomet ed. J. B. Hasted, Academic Press, London (1965).
2. N. F. Mott and H. S. W. Massey, *Theory of Atomic Collisions*, 3rd edition, Clarendon Press, Oxford (1965).
3. T.-Y. Wu and T. Ohmura, *Quantum Theory of Scattering*, Prentice Hall, Englewood Cliffs, N.J. (1962).
4. L. A. Vainshtein, I. I. Sobel'man and E. A. Yukov, *Electron-Excitation Cross Sections of Atoms and Ions* [in Russian], Nauka, Moscow (1973).
5. V. Ya. Veldre, in *Electron-Atom Collisions (Atomic Collisions, II)* [in Russian], Zinatne, Riga (1965), p. 3.
6. M. R. H. Rudge and M. J. Seaton, *Proc. Roy. Soc. A* 283, 262 (1965).
7. M. R. H. Rudge, *Rev. Mod. Phys.* 40, 564 (1968).
8. L. D. Landau and E. M. Lifshitz, *Quantum Mechanics* [in Russian], 2nd edition, Fizmatgiz, Moscow (1963); trans. by J. B. Sykes and J. S. Bell, Pergamon Press, London (1965).
9. R. K. Peterkop, *Proc. Phys. Soc.* 77, 1220 (1961).
10. A. Erdélyi, *Asymptotic Expansions*, Dover Publ., New York (1956).
11. É. Ya. Riekstyn'sh, *Latv. Matem. Ezhegodnik* 16, 88 (1975).
12. D. I. Blokhintsev, *Foundations of Quantum Mechanics* [in Russian], Vysshaya Shkola, Moscow (1961); trans: J. B. Sykes and M. J. Kearsley, *Quantum Mechanics*, Reidel, Dordrecht (1964).

13. E. Gerjuoy, *Ann. Phys.* 5, 58 (1958).
14. S. Cvejanović and F. H. Read, *J. Phys. B* 7, 1841 (1974).
15. H. Ehrhardt, K.-H. Hesselbacher, K. Jung and K. Willmann, *J. Phys. B* 5, 1559 (1972).
16. H. Ehrhardt, H.-H. Hesselbacher, K. Jung, M. Schultz and K. Willmann, *J. Phys. B* 5, 2107 (1972).
17. R. K. Peterkop, in *Atomic Collisions* [in Russian], Izd. Akad. Nauk Latv. SSR, Riga (1963), p. 115; trans. by M. V. Kurepa, Butterworth, London (1966), p. 107.
18. L. Castillejo, I. C. Percival and M. J. Seaton, *Proc. Roy. Soc. A* 254, 259 (1960).
19. R. K. Peterkop, *Opt. Spektrosk.* 13, 153 (1962); trans: *Opt. Spectrosc.* 13, 87 (1962).
20. R. K. Peterkop, in *Electron-Atom Collisions (Atomic Collisions II)* [in Russian], Zinatne, Riga (1965), p. 105.
21. R. K. Peterkop, *Izv. Akad. Nauk Latv. SSR,* No. 9, 79 (1960).
22. R. K. Peterkop, *Zh. Éksp. Teor. Fiz.* 43, 616 (1962); trans: *Sov. Phys.-JETP* 16, 442 (1963).
23. V. A. Fok, *Izv. Akad. Nauk SSSR, Ser. Fiz.* 18, 161 (1954).
24. A. M. Ermolaev, *Vestnik Leningrad Univ.* 22, 48 (1958).
25. Yu. N. Demkov and A. M. Ermolaev, *Zh. Éksp. Teor. Fiz.* 36, 896 (1959); trans: *Sov. Phys.-JETP* 9, 633 (1959).
26. A. Erdelyi, ed., *Higher Transcendental Functions,* Vols. I and II, McGraw-Hill, New York (1953).
27. D. W. Jepsen and J. O. Hirschfelder, *Proc. Nat. Acad. Sci.* 45, 249 (1959).
28. R. K. Peterkop and L. Rabik, *J. Phys. B* 5, 1823 (1972).
29. M. H. Hull and G. Breit, in *Handbuch der Physik,* Bd. 41, Springer-Verlag, Berlin (1959), p. 408.
30. V. V. Babikov, *Method of Phase Functions in Quantum Mechanics* [in Russian], Nauka, Moscow (1968).

31. R. K. Peterkop, *Izv. Akad. Nauk SSSR, Ser. Fiz.* 27, 1012 (1963).
32. J. D. Dollard, *J. Math. Phys.* 5, 729 (1964).
33. V. S. Buslaev and V. B. Matveev, *Teor. Mat. Fiz.* 2, 367 (1970); trans: *Theor. Math. Phys.* 2, 266 (1970).
34. A. M. Veselova, *Teor. Mat. Fiz.* 13, 368 (1972); trans: *Theor. Math. Phys.* 13, 1200 (1972).
35. L. A. Sakhnovich, *Teor. Mat. Fiz.* 13, 421 (1972); trans: *Theor. Math. Phys.* 13, 1239 (1972).
36. R. W. Hart, E. P. Gray and W. H. Guier, *Phys. Rev.* 108, 1512 (1957).
37. A. Temkin, *Phys. Rev. Letters* 16, 835 (1966).
38. R. K. Peterkop, in *Electron Scattering by Atoms (Atomic Collisions, IV)* [in Russian], Zinatne, Riga (1967), p. 35.
39. M. R. H. Rudge and M. J. Seaton, *Proc. Phys. Soc.* 83, 680 (1964).
40. R. K. Peterkop, in *Physics of the One- and Two-Electron Atoms*, eds., F. Bopp and H. Kleinpoppen, North Holland, Amsterdam (1969), p. 649.
41. G. Doolen and J. Nuttall, *J. Math. Phys.* 12, 2198 (1971).
42. Yu. A. Simonov, *Yad. Fiz.* 3, 630 (1966); trans: *Sov. J. Nucl. Phys.* 3, 461 (1966).
43. A. M. Badalyan and Yu. A. Simonov, *Yad. Fiz.* 3, 1032 (1966); trans: *Sov. J. Nucl. Phys.* 3, 137 (1966).
44. A. I. Baz' *et al.*, *Fiz. Élem. Chastits At. Yad.* 3, 275 (1972); trans: *Sov. J. Part. Nucl.* 3, 137 (1972).
45. L. M. Delves, *Nucl. Phys.* 9, 391 (1959); 20, 275 (1960).
46. W. Zickendraht, *Phys. Rev.* 159, 1448 (1967).
47. V. P. Zhigunov and B. N. Zakhar'ev, *Strongly Coupled Channel Methods in the Quantum Theory of Scattering* [in Russian], Atomizdat, Moscow (1974).

48. L. M. Delves, *Nucl. Phys.* 26, 136 (1961).
49. P. M. Morse and H. Feshbach, *Methods of Theoretical Physics*, Vol. 2, McGraw-Hill, New York (1953).
50. G. H. Wannier, *Phys. Rev.* 90, 817 (1953).
51. G. N. Lance, *Numerical Methods for High-Speed Computers*, Iliffe and Sons, London (1960).
52. M. Lieber, L. Rosenberg and L. Spruch, *Phys. Rev. D* 5, 1330, 1347 (1972).
53. R. K. Peterkop, in *Electron-Atom Collisions (Atomic Collisions, II)* [in Russian], Zinatne, Riga (1965), p. 139.
54. G. I. Kuznetsov, *Zh. Éksp. Teor. Fiz.* 51, 216 (1966); trans: *Sov. Phys.-JETP* 24, 145 (1967).
55. M. Bander and C. Itzykson, *Rev. Mod. Phys.* 38, 346 (1966).
56. R. K. Peterkop and M. P. Shkele, *Izv. Akad. Nauk Latv. SSR, Ser. Fiz. Tekhn. Nauk*, No. 2, 7 (1969).
57. R. K. Peterkop, in *Abstracts of VI Int. Conf. Phys. Electr. At. Coll.*, MIT Press, Cambridge, Mass. (1969), p. 936.
58. S. P. Merkur'ev, *Teor. Mat. Fiz.* 8, 235 (1971); trans: *Theor. Math. Phys.* 8, 798 (1971).
59. J. Nuttall, *J. Math. Phys.* 12, 1896 (1971).
60. W. Gordon, *Z. Phys.* 48, 180 (1928).
61. L. Marquez, *Am. J. Phys.* 40, 1420 (1972).
62. D. R. Yennie, D. G. Ravenhall and R. N. Wilson, *Phys. Rev.* 95, 500 (1954).
63. G. D. McCartor and J. Nuttall, *Phys. Rev. A* 4, 625 (1971).
64. S. P. Merkur'ev, *Teor. Mat. Fiz.* 17, 221 (1973); trans: *Theor. Math. Phys.* 17, 1105 (1974).
65. A. Dalgarno, *Phys. Rev.* 91, 198 (1958).
66. V. DeAlfaro and T. Regge, *Potential Scattering*, North Holland, Amsterdam (1965).
67. A. M. Lane and R. G. Thomas, *Theory of Nuclear Reactions at Low Energies*, IIL, Moscow (1960); this is a Russian

translation of "R-Matrix Theory of Nuclear Reactions," *Rev. Mod. Phys.* 30, 257 (1958).

68. S. Geltman, *Phys. Rev.* 102, 171 (1956).
69. W. H. Guier and R. W. Hart, *Phys. Rev.* 106, 297 (1957).
70. M. R. H. Rudge, *Proc. Phys. Soc.* 83, 419 (1964).
71. Yu. N. Demkov, *Variational Principles in the Theory of Collisions* [in Russian], Fizmatgiz, Moscow (1958); trans: N. Kemmer, Pergamon, Oxford (1963).
72. L. D. Faddeev, *Tr. Mat. Inst. Akad. Nauk SSSR* 69, 1 (1963).
73. V. A. Fok, *Zh. Éksp. Teor. Fiz.* 10, 961 (1940).
74. I. Zh. Vinkaln and M. K. Gailitis, in *Electron Scattering by Atoms (Atomic Collisions, IV)* [in Russian], Zinatne, Riga (1967), p. 17; trans. in: *Abstracts of V Int. Conf. Phys. Electr. At. Coll.*, Nauka, Leningrad (1967), p. 648.
75. L. D. Landau and E. M. Lifshitz, *Mechanics* [in Russian], Fizmatgiz, Moscow (1958); trans: J. B. Sykes and J. S. Bell, Pergamon, Oxford (1960).
76. G. H. Wannier, *Phys. Rev.* 100, 1180 (1955).
77. R. Abrines, I. C. Percival and N. A. Valentine, *Proc. Phys. Soc.* 89, 515 (1966).
78. V. F. Bratsev and V. I. Ochkur, *Zh. Éksp. Teor. Fiz.* 52, 955 (1967); trans: *Sov. Phys.-JETP* 25, 631 (1967).
79. R. K. Peterkop and P. B. Tsukerman, *Zh. Éksp. Teor. Fiz.* 58, 699 (1970); trans: *Sov. Phys.-JETP* 31, 374 (1970).
80. D. Banks, I. C. Percival and N. A. Valentine, in *Abstracts of VI Int. Conf. Phys. Electr. At. Coll.*, MIT Press, Cambridge, Mass (1969), p. 125.
81. D. Grujić, *J. Phys. B* 5, L137 (1972).
82. S. V. Khudyakov, *Zh. Éksp. Teor. Fiz.* 56, 938 (1969); trans: *Sov. Phys.-JETP* 29, 507 (1969).
83. R. Schiller, *Phys. Rev.* 125, 1100 (1962).

84. A. B. Migdal and V. P. Krainov, *Approximation Methods in Quantum Mechanics* [in Russian], Nauka, Moscow (1966); trans: A. J. Leggett, Benjamin, New York (1969).
85. A. K. Liepin'sh and R. K. Peterkop, *Izv. Akad. Nauk Latv. SSR, Ser. Fiz. Tekhn. Nauk*, No. 1, 17 (1969).
86. R. K. Peterkop and A. K. Liepin'sh, in *Abstracts of VI Int. Conf. Phys. Electr. At. Coll.*, MIT Press, Cambridge, Mass. (1969), p. 212.
87. R. K. Peterkop, *Izv. Akad. Nauk Latv. SSR, Ser. Fiz. Tekhn. Nauk*, No. 1, 7 (1971).
88. R. K. Peterkop, *J. Phys. B* 4, 513 (1971).
89. A. R. P. Rau, *Phys. Rev. A* 4, 207 (1971).
90. T. A. Roth, *Phys. Rev. A* 5, 476 (1972).
91. A. Temkin, A. K. Bhatia and E. Sullivan, *Phys. Rev.* 176, 80 (1968).
92. A. Temkin and Y. Hahn, *Phys. Rev. A* 9, 708 (1974).
93. C. E. Brion and G. E. Thomas, *Phys. Rev. Letters* 20, 241 (1968).
94. C. J. Krige, S. M. Gordon and P. C. Haarhoff, *Z. Naturforsch* 23a, 1383 (1968).
95. P. Marchand, C. Paquet and P. Marmet, *Phys. Rev.* 180, 123 (1969).
96. J. W. McGowan and E. M. Clarke, *Phys. Rev.* 167, 43 (1968).
97. I. P. Zapesochnyi and I. S. Aleksakhin, *Zh. Éksp. Teor. Fiz.* 55, 76 (1968); trans: *Sov. Phys.-JETP* 28, 41 (1969).
98. J. J. Thomson, *Phil. Mag.* 23, 449 (1912).
99. E. J. Williams, *Nature* 119, 489 (1927).
100. L. H. Thomas, *Proc. Camb. Phil. Soc.* 23, 713 (1927).
101. M. Gryzinski, *Phys. Rev.* 115, 374 (1959).
102. V. I. Ochkur and A. M. Petrun'kin, *Opt. Spektrosk.* 14, 457 (1963); trans: *Opt. Spectrosc.* 14, 245 (1963).

103. R. C. Stabler, *Phys. Rev.* 133, A1268 (1964).
104. L. A. Vainshtein *et al.*, *Zh. Éksp. Teor. Fiz.* 61, 511 (1971); trans: *Sov. Phys.-JETP* 34, 271 (1972).
105. V. I. Ochkur, in *Abstracts of VIII Int. Conf. Phys. Electr. At. Coll.*, Inst. Phys., Beograd (1973), p. 393.
106. L. Vriens, *Phys. Rev.* 141, 88 (1966).
107. M. Gryzinski, *Phys. Rev.* 138, A305, A322, A336 (1965).
108. A. E. Kingston, *Proc. Phys. Soc.* 87, 193 (1966).
109. A. E. Kingston, *J. Phys. B* 1, 559 (1968).
110. J. D. Garcia, *Phys. Rev.* 177, 223 (1969).
111. L. Vriens and T. F. M. Bonsen, *J. Phys. B* 1, 1123 (1968).
112. D. Banks, L. Vriens and T. F. M. Bonsen, *J. Phys. B* 2, 976 (1969).
113. K. L. Bell, M. W. Freeston and A. E. Kingston, *J. Phys. B* 3, 959 (1970).
114. L. Vriens, *Physica* 47, 267 (1970).
115. D. R. Bates and W. R. McDonough, *J. Phys. B* 3, L83 (1970); 5, L107 (1972).
116. M. Inokuti, *Rev. Mod. Phys.* 43, 297 (1971).
117. L. A. Vainshtein, *Opt. Spektrosk.* 3, 313 (1957).
118. G. Peach, *Proc. Phys. Soc.* 85, 709 (1965); 87, 375, 381 (1966).
119. G. Peach, *J. Phys. B* 1, 1088 (1968).
120. G. Peach, *J. Phys. B* 3, 328 (1970).
121. G. Peach, *J. Phys. B* 4, 1670 (1971).
122. H. L. Kyle and K. Omidvar, *Phys. Rev.* 176, 164 (1968).
123. K. Omidvar, H. L. Kyle and E. C. Sullivan, *Phys. Rev. A* 5, 1174 (1972).
124. E. J. McGuire, *Phys. Rev. A* 3, 267 (1971).
125. H. Bethe, *Ann. Phys.* 5, 325 (1930).
126. M. Inokuti and Y. K. Kim, *Phys. Rev.* 186, 100 (1969).

127. Y. K. Kim and M. Inokuti, *Phys. Rev. A* 3, 665 (1971).
128. K. Omidvar, *Phys. Rev.* 177, 212 (1969).
129. K. L. Bell and A. E. Kingston, *J. Phys. B* 2, 1125 (1969).
130. I. I. Sobel'man, *Introduction to the Theory of Atomic Spectra* [in Russian], Fizmatgiz, Moscow (1963).
131. V. V. Balashov, S. S. Lipovetskii and V. S. Senashenko, *Vestn. Mosk. Univ., Fiz. Astronomiya,* No. 1, 116 (1973).
132. W. K. Peterson, C. B. Opal and E. C. Beaty, *Phys. Rev. A* 5, 712 (1972).
133. E. Stingl, *J. Phys. B* 5, 1688 (1972).
134. E. Stingl, *J. Phys. B* 5, 1160 (1972).
135. R. J. Tweed, *J. Phys. B* 6, 259, 270, 398 (1973).
136. R. K. Peterkop, *Zh. Éksp. Teor. Fiz.* 41, 1938 (1961); trans: *Sov. Phys.-JETP* 14, 1377 (1962).
137. S. Geltman, M. R. H. Rudge and M. J. Seaton, *Proc. Phys. Soc.* 81, 375 (1963).
138. I. H. Sloan, *Proc. Phys. Soc.* 85, 435 (1965).
139. M. R. H. Rudge and S. B. Schwartz, *Proc. Phys. Soc.* 88, 563, 579 (1966).
140. S. S. Prasad, *Proc. Phys. Soc.* 87, 393 (1966).
141. W. L. Fite and R. T. Brackmann, *Phys. Rev.* 112, 1141 (1958).
142. V. I. Ochkur, *Zh. Éksp. Teor. Fiz.* 45, 734 (1963); trans: *Sov. Phys.-JETP* 18, 503 (1964).
143. V. I. Ochkur, *Zh. Éksp. Teor. Fiz.* 47, 1746 (1964); trans: *Sov. Phys.-JETP* 20, 1175 (1965).
144. V. I. Ochkur, in *Abstracts of VI Int. Conf. Phys. Electr. At. Coll.,* MIT Press, Cambridge, Mass. (1969), p. 251.
145. L. Vainshtein, L. Presnyakov and I. Sobel'man, *Zh. Eksp. Teor. Fiz.* 45, 2015 (1963); trans: *Sov. Phys.-JETP* 18, 1383 (1964).

146. L. Presnyakov, *Zh. Éksp. Teor. Fiz.* 47, 1134 (1964); trans: *Sov. Phys.-JETP* 20, 760 (1965).

147. L. Presnyakov, *Tr. Fiz. Inst. Akad. Nauk SSSR* 51, 20 (1970).

148. D. S. F. Crothers, *Proc. Phys. Soc.* 91, 855 (1967).

149. V. Ya. Veldre and I. Zh. Vinkaln, *Opt. Spektrosk.* 18, 902 (1965); trans: *Opt. Spectrosc.* 18, 507 (1965).

150. I. Zh. Vinkaln, in *Electron-Atom Collisions (Atomic Collisions, II)* [in Russian], Zinatne, Riga (1965), pp. 87, 97.

151. L. A. Vainshtein, *Opt. Spektrosk.* 11, 301 (1961); trans: *Opt. Spectrosc.* 11, 163 (1961).

152. R. K. Peterkop, *Opt. Spektrosk.* 12, 145 (1962); trans: *Opt. Spectrosc.* 12, 77 (1962).

153. A. K. Leipin'sh, *Izv. Akad. Nauk Latv. SSR, Ser. Fiz. Tekhn. Nauk*, No. 6, 22 (1968).

154. A. K. Liepin'sh, *Izv. Akad. Nauk Latv. SSR, Ser. Fiz. Tekhn. Nauk*, No. 2, 3 (1970).

155. A. K. Liepin'sh and L. L. Rabik, *Izv. Akad. Nauk Latv. SSR, Ser. Fiz. Tekhn. Nauk*, No. 6, 34 (1972).

156. P. G. Burke and A. J. Taylor, *Proc. Roy. Soc. A* 287, 105 (1965).

157. M. B. Hidalgo, J. H. McGuire and G. D. Doolen, *J. Phys. B* 5, L70 (1972).

158. J. H. McGuire, M. B. Hidalgo, G. D. Doolen and J. Nuttall, *Phys. Rev. A* 7, 973 (1973).

159. D. F. Gallaher, *J. Phys. B* 7, 362 (1974).

160. R. Damburg and E. Karule, *Proc. Phys. Soc.* 90, 637 (1967).

161. S. S. Lipovetskii, A. V. Pavlichenkov, A. N. Polyudov and V. S. Senashenko, *Vestn. Mosk. Univ., Fiz. Astronomiya*, No. 4, 452 (1971).

162. V. V. Balashov, S. S. Lipovetskii, A. V. Pavlichenkov, A. N. Polyudov and V. S. Senashenko, *Opt. Spektrosk.* 32, 10 (1972); trans: *Opt. Spectrosc.* 32, 4 (1972).

163. V. V. Balashov, S. S. Lipovetskii and V. S. Senashenko, *Zh. Éksp. Teor. Fiz.* 63, 1622 (1972); trans: *Sov. Phys.-JETP* 36, 858 (1972).

164. S. S. Lipovetskii and V. S. Senashenko, *Opt. Spektrosk.* 34, 1046 (1973); trans: *Opt. Spectrosc.* 34, 607 (1973).

165. A. V. Pavlichenkov and V. S. Senashenko, *J. Phys. B* 5, 1898 (1972).

166. V. V. Balashov, S. S. Lipovetskii and V. S. Senashenko, *Vestn. Mosk. Univ., Fiz. Astronomiya*, No. 4, 503 (1973).

167. V. V. Balashov, S. I. Grishanova, I. M. Kruglova and V. S. Senashenko, *Opt. Spektrosk.* 28, 859 (1970); trans: *Opt. Spectrosc.* 28, 466 (1970).

168. V. V. Balashov, S. S. Lipovetskii, A. V. Pavlichenkov, A. N. Polyudov and V. S. Senashenko, *Vestn. Mosk. Univ., Fiz. Astronomiya,* No. 1, 65 (1971).

169. U. Fano, *Phys. Rev.* 124, 1866 (1961).

170. A. K. Liepin'sh and L. L. Rabik, *Izv. Akad. Nauk Latv. SSR, Ser. Fiz. Tekhn. Nauk,* No. 6, 24 (1972).

171. A. K. Liepin'sh and L. L. Rabik, *Izv. Akad. Nauk Latv. SSR, Ser. Fiz. Tekhn. Nauk,* No. 6, 42 (1972).

172. Yu. P. Korchevoi and A. N. Przhonskii, *Zh. Éksp. Teor. Fiz.* 51, 1617 (1966); trans: *Sov. Phys.-JETP* 24, 1089 (1967).

173. K. J. Nygaard, *J. Chem. Phys.* 49, 1995 (1968).

174. W. M. Bryce and F. Mandl, *J. Phys. B* 5, 912 (1972).

175. C. Bottcher, W. M. Bryce and F. Mandl, *J. Phys. B* 7, 769 (1974).

176. L. Rosenberg, *Phys. Rev. D* 8, 1833 (1973).

177. L. L. Rabik, in *Atomic Processes* [in Russian], Zinatne, Riga (1975), p. 25.

178. U. Fano, *J. Phys. B* 7, L401 (1974).

179. S. Cvejanović and P. Grujić, *J. Phys. B* 8, L305 (1975).

180. D. Spence, *Phys. Rev. A* 11, 1539 (1975).

181. W. D. Robb, S. P. Rountree and T. Burnett, *Phys. Rev. A* 11, 1193 (1975).

182. A. Jain and M. K. Srivastava, *J. Phys. B* 9, 1103 (1976).

183. K. J. Nygaard, *Phys. Rev. A* 11, 1475 (1975).

184. S. N. Tiwary and D. K. Rai, *J. Phys. B* 8, 1109 (1975).